INVENT

Yourself Rich

INVENT

16 Secrets for Creating Million-Dollar Inventions

Yourself Rich

Don Brown

Foreword by Don Hall

WESTHOLME
Yardley

Westholme Publishing, LLC
Eight Harvey Avenue
Yardley, Pennsylvania 19067
Visit our Web site at www.westholmepublishing.com

First Edition: April 2007
10 9 8 7 6 5 4 3 2 1

ISBN: 978-1-59416-050-9
ISBN 10: 1-59416-050-3

Printed in United States of America

This book is dedicated to YOU the Inventor

Thank you for following your spark of innovation and seeing it through. Without people like you, exciting new products would never become a reality.

Contents

Foreword

No, I don't write forewords. . . .

No, I don't have the time to write a foreword. . . .

No, it's not possible. . . .

OK, I'll take a look at it. . . .

OK, I'll read it on my next flight. . . .

YES, the book is the real deal. . . .

YES, I'll write a foreword.

So was the journey that led me to scribbling these words at the front of this practical handbook of proven techniques for success with inventions.

My conversion is testimony to Don's Secret #6: Never Give Up!

To be OVERT—Don Brown has written THE BOOK for those interested in increasing their odds of invention success and decreasing their chances of costly pain. Read it—live it—or don't whine about how you're not having success.

WARNING: Don has told the truth, the whole truth and nothing but the truth. His lessons are NON-NEGOTIABLE.

When Don says, "The best way to get an honest answer is to have actual samples (working prototypes) to show and demonstrate," he means it.

When Don says, "The 5:1 ratio is an ironclad rule of retail," he means it.

When Don says, "Try not to be too emotionally attached to your invention," he means it.

When Don says, "97 percent of all patents never make a profit!" he means it.

When Don says, "Focus on selling, not suing," he means it.

When Don says, "They will invest in facts, not hype," he means it.

When Don says, "Companies don't care about ideas—they care about products," he means it.

It takes vision and courage to be an inventor. Inventor Don Brown exposes the nine scams that each year cost hundreds of thousands of inventors millions of dollars.

Now if I could only get the contestants on ABC-TV's *American Inventor* to read Don's book AND heed his advice, I wouldn't get nearly as frustrated each week.

Why Don's book is of critical importance to the United States: Our country, our planet is in need of inventors. From global warming to global dependence on oil—from diabetes to obesity—from the decline in high school math scores—to our falling to 16th in the world in our levels of new scientists and engineers—we desperately need inventors to help solve the world's problems.

Don provides the ideas, insights, and inspiration you need to invent meaningful solutions to the challenges of our community, state, nation, and planet. To be clear—if your idea is a BIG IDEA—a BIG, BOLD, MARKET-CHANGING IDEA—you will cause FRICTION!

All BIG IDEAS cause friction. It's just a fact of life. Hold fast. Stay the course. In time, innovation is usually considered more right than wrong.

Consider this 1829 letter from Martin Van Buren, then governor of New York, to President Andrew Jackson regarding the dangerous risk that the innovation of railroads posed to the state of New York.

> To: President Jackson
>
> The canal system of this country is being threatened by the spread of a new form of transportation known as "railroads." The federal government must preserve the canals for the following reasons:
>
> ONE—If canal boats are supplanted by "railroads," serious unemployment will result. Captains, cooks, drivers, hostlers, repairmen and lock tenders will be left without means of livelihood, not to mention the numerous now employed in growing hay for horses.

> TWO—Boat builders would suffer and towline, whip and harness makers would be left destitute.
>
> THREE—Canal boats are absolutely essential to the defense of the United States. In the event of expected trouble with England, the Erie Canal would be the only means by which we could ever move the supplies so vital to waging modern war.
>
> As you may well know, Mr. President, "railroad" carriages are pulled at the enormous speed of 15 miles per hour by "engines" which, in addition to endangering life and limb of passengers, roar and snort their way through the countryside, setting fire to crops, scaring livestock and frightening women and children. The Almighty certainly never intended that people should travel at such breakneck speed.
>
> Signed
>
> Martin Van Buren, Governor of New York

Van Buren's fear of the impact of railroads is today seen as funny. So too will resistance to your bold and brave ideas be seen as humorous in the future.

Our world faces some serious challenges. Pessimism is at epidemic proportions. Myself—I'm optimistic—optimistic because of the American

Inventors and people like Don Brown who graciously share their wisdom so that all of us can be smarter.

So what are you waiting for—start reading and applying Don's wisdom—TODAY!

—Doug Hall

Introduction

One good idea could make you a millionaire.

One day while working as a fitness trainer in a health club I had an idea for an invention that would make sit-up exercises a little easier. I decided to take action and turn my idea into a product. With a piece of pipe, some duct tape, and a cushion, I made a working model. The product worked perfectly, and two years later I was a millionaire. More than 10 million people own Ab Rollers today, and it continues to sell in stores all over the world.

I never went to college, I am not an engineer, and my family never had any money. I had a good idea, I took action, and I INVENTED MYSELF RICH—and so can you. All it takes is one good idea. More and more people are becoming millionaires by creating a product that millions of consumers are willing to purchase. Think about how easy it would have been for you to invent the Pasta Pot. This product did not require a degree in engineering. All the inventor had to do was drill a few holes in the lid of his favorite pot, file a provisional patent, and license it to a TV infomercial company. You could have invented it and made a small fortune.

You probably already have a good idea about what you want to invent, but do you know what to do next? Most people don't. That's why I wrote this book—to share my knowledge with inventors like you in the hope of shedding much-needed light on the business of inventing.

In this book, you will learn how to come up with good ideas, how to test your ideas, and when to invest your time and money. You will also learn valuable lessons on what not to do. You will learn how to protect your idea from unscrupulous invention companies, attorneys, and patent agents out there trying to make money off your idea before you make a penny. I'll be sharing sixteen secrets for successful inventions and nine tips to avoid being scammed along the way. This book will save you thousands of dollars by avoiding foolish mistakes on unnecessary patent services, trademarks, or focus groups. If after reading this book you discover that your idea isn't a good one after all, you will have saved time and money. You can always apply the information you gain in this book for your next idea.

The stories in this book are about average people just like you who had a good idea, a little passion, and the desire to see it through. They created a small fortune by using the secrets you are about to discover. You may never win the lottery or inherit millions from your family, but you are about to discover how you can INVENT YOURSELF RICH!

Part One

Chapter One

The Opportunity for Invention

"Everyone has a Million-Dollar Idea. It's knowing what to do with it that will make you a millionaire."—Don Brown

If I Build It, Will They Come?

There are more than three hundred million people living in the USA, just waiting for the next revolutionary and remarkable invention to hit the market. Not only do consumers want new and exciting products, big companies like Procter and Gamble, Hasbro, Dial Corporation, and Crayola also want the next big hit, and they are looking outside the four walls of their company for inventions. Did you know that the ideas for 90 percent of the toys manufactured by major companies come from outside the company? But wait—there's more! The direct-response infomercial companies also want your products. Home Shopping Network, Shop at Home, Value Vision, and QVC thrive off inventors, and they wouldn't have a business without the latest new-fangled gadget or game inventors provide them. The quest for invention is evident with the launch of the TV reality show *American Inventor.*

As you can see, the opportunity to invent yourself rich is out there waiting for you. All you have to do is think of a need or problem and come up with a product to solve it in a manner that is better than anything currently available.

Inventing yourself rich will change your life forever. Doshu Shifferaw came to southern California from Ethiopia in 1973 as a college student to study political science. Soon after his arrival, his education plans were thwarted when revolution broke out in Ethiopia. His father, who was funding his tuition, was put in jail, and Doshu had to drop out of college. He often dreamed about inventions while driving cabs and working at hotels in San Francisco to make ends meet. When he shared his invention ideas, people told him his inspiration was silly. Some tried to get him to give them his design plan. Instead, he sat on the beach and separated himself from their negativity and perfected his sketches. Doshu had a vision and just knew his product would succeed. He was right. His idea emerged into an exercise chair that he patented and sold around the world as Bowflex, licensed by The Nautilus Group. Not only did Doshu invent the Bowflex, it invented him. It changed his life, as well as the life of his family and relatives. No more driving cabs and hoping for a big tip. He invented himself rich, and so can you!

When you have a great idea, you are filled with excitement. You may not be sure what to do first. Contrary to popular thought among amateur inventors, securing a patent is not the first thing you should do. In fact, this is the single biggest mistake inventors make. The main reason inventors spend money on a so-called "patent expert" is because they mistakenly believe their idea should be patented. Why? Because they are fearful of someone ripping them off, and they believe a patent will protect them. Allowing this fear to dictate your actions in the early stages is a huge mistake. A patent applies only to inventions that fall within legally defined categories. Anything that lies outside these categories cannot be patented. I will explain more about the patent in Chapter 2.

Lawyers can draft a patent and put a contract together, but they have no idea whether a product has a chance of succeeding. Your instincts are as good as theirs, and you can do the research to determine marketability for yourself. My own experience proves that much. I was a meat cutter (butcher) for nine years. I never went to college, but I took classes at night to learn about fitness and nutrition. Then, I worked as a personal trainer and knew the fitness industry quite well. When I first invented the Ab Roller, I took my invention to a patent lawyer. He helped me file the patent application. Yes, I did file a patent for the Ab Roller, but not until I had a tested working prototype. Back then, a provisional patent was not an option. Today a provisional patent is the way to go before you publicly disclose your invention.

Then, I took the product to Nordic Track, a well-known consumer fitness company. Nordic Track offered me $25,000 plus a 4% royalty, and my lawyer thought it was a fair offer and suggested that I take it. Fortunately, I received better advice from others and didn't listen to him. I had received very positive feedback on my product and felt that I had something really good. I never dreamed that worldwide sales of my invention would exceed $750 million! As you can see, you do not need a patent attorney at the beginning of the invention process. You may need a lawyer later in the process; however, I will discuss hiring a patent agent or attorney in Chapter 2.

Patenting costs a small fortune. Therefore, my philosophy on the invention business is very simple. Once you have a great idea, you should work on turning your idea into a real and marketable product. Rather than spend $10,000 on a patent that may never make you any money, I urge inventors to spend money on developing a product they can test in the marketplace, before spending money on a full patent. After all, why patent an invention that may never be sold? Especially since the U.S. Patent and Trademark Office reports that 97 percent of all patents never make a profit! You don't want to be part of the 97 percent.

Thomas Edison once said, "I don't want to invent anything that won't sell!" I have to agree. So, if you plan to invent something that will make you rich, you need to invent something that millions of people will either want or need. Otherwise, your product will not sell well enough to make a profit or you may end up losing money and wasting your time.

Let's assume you have an idea. It may be a good idea, but the next thing you need to ask yourself is whether or not your idea is a perfect idea for a perfect product.

What Is the Perfect Product?

All million-dollar products begin with an extraordinary idea. In *Purple Cow*, a book about business marketing, Seth Godin describes unique products and services as "remarkable" or worth talking about—a product that not only delivers what customers expect but does so much more. He says: "Cows, after you've seen them for a while, are boring. They may be perfect cows, attractive cows, cows with great personalities, cows lit by beautiful light, but they're still boring. A Purple Cow, though, now that would be interesting. (For a while.)" (p. 2).

Coming up with an idea (even if it is remarkable) is not enough. In order for a remarkable or perfect idea to become an invention, it must be made to work. The perfect idea must also convert into a product that can be produced at a price that makes a profit while remaining competitive in the marketplace—even if there are similar products already on the market. I will discuss the ideal ratio of retail price to manufactured cost later in this book.

To make sure you do not publicly disclose your idea in the early stages, you should discuss your idea only with your family and with friends you know best and trust most. The best way to get an honest answer is to have actual samples of the product (working prototypes) to show and demonstrate. At the beginning stage, however, to avoid the cost of creating an actual prototype, you can use a simple drawing. Even an oral or written description can be used. If you can't get a "yes" from someone, anyone, even from your own mother, then maybe you should focus on another idea.

American Inventor is a fast-paced television series that embodies the American dream of becoming rich, famous, or successful overnight—in this case through a clever invention. Potential inventors present their invention to an expert panel. Through a process of elimination, the winning invention is awarded one million dollars to help the inventor to turn his or her idea into a mass-produced product that can be sold at a profit.

The reality show's success is proof that people believe they can do anything if the funds are available. Even the winning invention must be accepted by the public and manufactured and sold at a profit before it can be considered a success. That means a product must meet certain criteria before it is considered a perfect product.

Secret #1: Inventing the Perfect Product

A perfect product fills a common need or answers a perplexing question in a unique or better way than anything on the market. It can be easily manufactured or incorporated into an existing product. There will be a market large enough to support the product in its production and turn a profit as soon as it hits the market. The perfect product will cause the market to grow in spite of any competition. Take the Java Jacket as a case in point.

Have you ever seen a hot new product on TV or at the store and said, "I've thought of that idea. I should have invented it myself!" Prior to creating my first invention, I experienced a lot of these moments when I saw the latest fitness device on TV, but the best product ever to slip by me was when I was in the airport about twenty years ago. My friend who was traveling with me asked me to get her a cup of coffee. She also asked me to get her two cups because she didn't like holding a hot cup. I thought to myself, "Using two cups is such a waste!" I went home and took a small foam cup and sliced the top one inch off and slid it up on a cup. Voilà—a coffee cup sleeve to protect her hand from burning. What a great idea! I eliminated the need to use a second whole cup, and avoided all that waste. This viable product could be manufactured for pennies and easily sold at a profit to any store that sells hot drinks.

So, guess what I did next? Absolutely nothing. A few years later I was in a coffee shop and they handed me a hot cup of coffee in a Java Jacket invented by Jay Sorensen. Jay had been holding a cup of coffee in his hand when the paper cup he was using became too hot to hold. Instinctively he let go and it fell into his lap. His heated experience gave him inspiration for his invention. He quickly turned his house into a factory where he created the Java Jacket—an insulated sleeve that fits a coffee cup. It is made of earth-friendly recycled paper and may be printed with a company logo that lets customers advertise the vendor's business while they drink their coffee. He started selling Java Jackets to local coffee shops using the trunk of his car as a retail venue. When he took his product to a coffee trade show in Seattle, it became an immediate success nationally. Over one billion cup sleeves have been sold. The last report I read stated the company who makes this simple little product was doing over $8 million a year in sales!

While I was holding my first cup of coffee sporting a Java Jacket I remembered that day in the airport, and how I had thought of the idea which now burned in my mind. OOPS! I MISSED THAT ONE. Don't you do the same! The moral of this story is two-fold. First, the Java Jacket is an example of the perfect product. Second, don't miss your million-dollar idea through procrastination—act immediately.

Patent It First?

I THINK NOT! Let's suppose you have a perfect idea and you are going to take the next step to bring it to market. You probably think you should apply for a patent first thing, right? Wrong. You should not rush out immediately and secure a patent for your product. Many invention promoters, and even patent attorneys, will try to persuade you to apply for a patent on your idea before you do anything else. They will warn you about losing your rights or being ripped off if you do not obtain a patent first. However, you should think twice about spending thousands of dollars on a full patent before you know that your invention can be sold successfully. Instead, you should get a provisional patent before you disclose your idea publicly. Your dollars should be spent initially on developing an actual product from your idea to learn whether it is truly dramatically different and whether your invention can be successfully marketed. There are ways to protect your idea while you develop it, and I will discuss those ideas later. For now, you need to start your own research on what is already out there.

Search and Research

There are two reasons you need to conduct a prior art search (existing knowledge) and a patent search before moving any further with your perfect idea. First, the search will help distinguish your invention from other items on the market and will strengthen your patent application—if you decide you need one. If you find there is something already patented or on the market too similar to the idea you have conceived, it may be wise to move on to another idea before spending any money on a product that will not invent riches for you. And second, it will keep you from infringing upon someone else's patent. Inventors are required by law to submit their findings to the U.S. Patent and Trademark Office when making application for a patent. If it is discovered that you knew about a similar innovation and did not report that information to the Patent Office, you can be sued for fraud.

Before you hire a patent attorney, there are things you can do on your own to find out whether or not you should continue with your invention. At this point, you are trying to determine marketability as well as searching for patents, so I advise inventors to start by deciding whether there is something similar already on the market. This may seem like an obvious step, but you would be amazed at how many people forget to do this. There are many free ways to determine whether your idea is already on the market. The easiest place to start is in your local Wal-Mart or other large retail store. Scan the shelves and various departments for your idea, and see if it already exists. You would be surprised at how many products already exist that you may have never heard about. This is often due to a lack of marketing or ineffective marketing that causes products to languish on the shelves because they are not packaged or promoted properly. Also check the Internet by performing a search on www.google.com or www.yahoo.com. Type in a word or word phrase that describes your invention and see if you find something similar.

If you do find items on the market that are similar to yours, you need to be honest with yourself and ask yourself these questions:

Will people want to buy and use my product instead of the competitor's?

Is my product superior to the competition?

Why should the customer change what they are using now?

Is my idea noticeably different from what is currently available?

There are many ideas that have been patented but have never come to market (remember the 97 percent statistic I mentioned earlier). So be aware that, even though your idea may not be in a store, someone may have thought about your idea already, and they may have spent a lot of money securing a patent for it. You need to confirm that your idea does not infringe upon another person's patent rights. To do this, you can go to the Patent Office and search it yourself, or you can go online and review all the patents on the government's site (www.uspto.gov). I suggest you make use of online databases such as the Internet Patent Search System, Source Translation and Optimization (STO), The Public Search Room of the U.S. Patent and Trademark Office (USPTO), the Delphion Research Intellectual Property Network (formerly IBM patenting network) the Thomas Register, Dun & Bradstreet's Directory, and Standard & Poor's Register. Also search academic and technical literature such as trade magazines and directories and your local patent depository library for mention of ideas similar to yours.

To be completely sure, you could hire a patent agent (or attorney) to conduct a patent search. This search should not cost more than $1,000. The patent search is only to see if there is already patent for the product you want to make. It does not protect your invention in any way. To protect your invention you need to file a provisional patent after you have a working model of your product. You will then have one year to perfect the product and apply for a full patent. The median reported by U.S. patent attorneys for the year 2002 was $5,504 in fees for services, plus actual out-of-pocket costs such as the government filing fee, to prepare and file a full U.S. utility patent application on an invention of "minimal complexity." This number is from the Economic Survey published by the American Intellectual Property Law Association in 2003—a survey conducted every two years. An attorney will charge additional time to shepherd the patent application all the way through issuance by the Patent Office. Since these fees are from 2003, the fees and costs are somewhat higher now. Fees depend on the experience and skill of the patent practitioner, the complexity of the invention, the number of claims, the number of patent applications, and other factors.

If your product is not noticeably different, or if it infringes upon another patent, it's time to think of another new idea.

In order to determine whether or not your idea will qualify as a perfect product, you also need to do some market research.

Every year, new products are introduced into the marketplace but only a few succeed in a big way. Several products are new versions of existing products designed to keep the category fresh and selling. But a few new products or what I call "perfect products" will explode and create millions of dollars in sales. When my Ab Roller hit the market, people instantly understood the problem it was designed to solve and how it functioned to resolve the problem. Americans want a lean midsection and they are willing to try anything that will work. Because the Ab Roller really did as it claimed, millions of people actually felt their abdominal muscles working and began spreading the word. More than 10 million people own Ab Rollers, and sales continue year after year. The Ab Roller has become its own product category.

If you do not wish to do your own market research, there are professional firms which provide this service for a fee. However, they are very costly, and I would not recommend spending that much in the beginning. To get low-cost help in market research, you might want to contact an entrepreneurial business faculty member or group at a local college. Students are often willing to take on such projects. The Small Business Administration (SBA) or local government agencies may be a good resource for finding links to the business and academic communities. National and local inventors clubs, and retired business men and women may provide business perspectives and marketing information and advice for little or no money.

How Do You Create a Perfect Product?

Have you ever had a day when everything goes better than expected? If you are a golfer perhaps you hit the ball perfect all day long and scored better than par. Gymnast Mary Lou Retton had a perfect day when she scored a 10 on the vault and became the first American woman to win an individual medal in Olympic gymnastics.

The perfect product is very similar—it has all the components necessary for consumer acceptance: a large market, a trend in a particular category, cost to retail price ratio, the right name, the right package, the right marketing message—strong patent and trademark. My product was almost a perfect product. The Ab Roller was not the name I originally gave my product. I named it the Ab Trainer. Luckily I was able to obtain ownership of the Ab Roller name as part of the settlement when someone infringed upon my patent.

Think of how you feel when you eat healthy, sleep well, and are happy with your life. You live longer and feel better. That is what a perfect product can do. It can give people more enjoyment in life and outlast other trends in the market.

Secret #2: Catch a Trend

I don't know who actually came up with the idea of putting a yellow ribbon on a car in support of the

U.S. troops in Iraq, but Magnet America is the company that produces them. I doubt the original inventor filed a patent for the "Support our troops" ribbon. The inventor didn't have time to wait out the legal process (who can know how long a war will last?). Instead, the inventor forged ahead and grabbed the dollar while the trend was hot. Likewise, Lance Armstrong created the Live Strong yellow wristband and sold them for a dollar a piece to raise money for his cancer foundation. The emotional wave of public opinion that followed created a million-dollar fad. Who knows if the Lance Armstrong Bracelet would have been profitable without the desire to donate for cancer research, but linking a product to a trend or charity can be profitable for the inventor who is first out and best dressed.

Many imitations will follow a fresh idea, so if you find items already on the market that are similar to your idea, it does not mean you should abandon your invention. The magnetic ribbon or wristband that expresses an opinion is something that could easily be copied using a different color, size, or message. Since religious and patriotic merchandize has such a wide appeal, these ribbons and wristbands continue to find commercial success. Just think of all the different magnetic ribbons you see these days. Nearly every car on the road has a ribbon with a message that represents the driver's political stance or support for a cause.

Don't waste money on a patent that will be hard to protect. Catch a trend, ride a wave, invent yourself rich!

Secret #3: Invent a Trend Through Collectibles

Just as successful as the product being copied in multiple designs is the product that is produced in minimum quantities to give the illusion of scarcity. Take the Beanie Babies as an example. Introduced by H. Ty Warner's Ty Toy of Illinois in 1993, these small understuffed animals were affordably priced and had cute names that kids love. The original nine—Squealer the pig, Spot the dog, Chocolate the moose, Patti the platypus, Cubbie the bear, Pinchers the lobster, Splash the killer whale, Legs the frog, and Flash the dolphin—took Chicago by storm in 1994. Soon the adults were in on it. Fights broke out during the holiday shopping season when people tried to grab the limited stock at the few retailers who sold them.

Warner not only limited the supply, he also limited his availability to the media. He rarely grants interviews. Ty Incorporated is a private company located in Illinois and has an unlisted telephone number and offers only one hundred shares of stock (Ty Warner possesses all of them). The limited edition approach made the Beanie Babies a highly sought-after collectible and Ty Warner a multibillionaire. The Beanie kingdom is now wealthier than Hasbro and Mattel combined. That's pretty good success for a guy who was a college drop-out.

Warner is still birthing Beanie Babies and collectors are still buying his commemorative releases. Ty Rescue the Dog—Beanie Baby 9-11 was released in honor of the 9/11 fire and police rescuers. Showing his philanthropist side, Warner contributed 100 percent of the profits to the New York Police & Fire Widows' & Children's Benefit Fund. These stuffed reminders of the tragic event could be purchased for only $6.99, but of course there was a limit of two pups per order. Interestingly, in contrast to his limited edition strategy, Ty has allowed miniature Beanie Babies to be licensed to fast-food chains. Kids everywhere could find a McDonald's Beanie in their Happy Meal box.

The Cabbage Patch Kids (originally called Little People) are another example of a scarcity collection phenomenon. After being rejected by Mattel Corporation, Zavier Roberts licensed the Little People to Coleco Industries, which manufactured them with vinyl heads and renamed them the Cabbage Patch Kids. The big selling point for these dolls was the adoption gimmick associated with them. The story on the box tells about a boy named Xavier who found some special Little People in a cabbage patch who wanted to have families adopt them and share their love. Each doll came with adoption papers, a birth certificate, and an adoption oath for the adoptive "parent" to abide by. You could send in the adoption registration form to Coleco and a certificate of adoption would be sent as well as a birthday card on the Kid's birthday the next year. The 16-inch, soft-bodied dolls "born" in Babyland General Hospital in Cleveland, Georgia, were normally priced at $25 to $30 each. They were in such high demand that they brought more than $300 each on the black market during the holiday season in 1983.

Coleco filed for bankruptcy in 1988. You wonder how that could happen after having selling 3 million CPK dolls and 20 million Preemies and Koosas dolls. They were selling more CPKs than Barbie dolls at one time, but the demand decreased and Coleco started trying to sell talking and burping dolls, which didn't do well at all. That's when Hasbro bought Coleco and continued with posable dolls that could blow a kazoo. In 1994, Mattel took over the license for Cabbage Patch Kids, but consumer interest fell off significantly until the fifteenth anniversary Kid was released in 1998.

One of the most recent and successful booms in collectibles is the State Quarters program. When the United States Mint decided to launch the 50 State Quarters program, who would have thought that it would create a 120 million-dollar collection craze? A few very smart and savvy inventors and TV marketers took action and created the US Coin Maps. These brightly colored cardboard foldout coin maps designed to hold a quarter in each state cost less than $3.00 a unit and were selling like crazy for $19.95! One company in New Jersey sold over 6 million maps within the first 6 months. They even presold coins for the next 4 years promising the consumer that the coins would be in semi-mint condition. The customer was willing to pay over $1.50 per 25 cent coin to have someone else pick it out of a bag and mail it to them. Who would of thought you could sell money for more than it was worth?

You never know when your product may become a collectible, but you might have that end in mind when you are thinking of toys. Trends come in all shapes and sizes. Your trend product may stay on the market for years by reintroducing them with a small modification such as the Cabbage Patch dolls. Even though the license has changed hands several times, Xavier Roberts is still making money on his invention.

Secret #4: Classic Toys Never Die

You can take a top, a game, set of jacks, or a baby doll, reinvent it, and make a fortune. The spinning top is probably one of the most recognized classic toys ever created. Yet each year someone reinvents it into a new exciting product.

One of my favorite reinventions is the I-top from Irwin toys. They have taken the classic spinning top and added a minicomputer inside that controls a row of LED lights. With it, you can play a number of games, such as "Spin to Win." When the top is spun, LEDs light up and give you the illusion of animated text and numbers displaying the number of revolutions the top spins before coming to a complete stop. Players try to obtain the highest number of rotations. Now that is cool!

What product could you think of that would invent yourself rich? A perfect idea is one that is so different from what is presently available it stands out from the crowd and commands attention. Breathe Right Strips, Liquid Paper, and Velcro are examples of perfect products. Let's look at the secret behind the success of these inventions.

Secret #5: Invent a Consumable Product

A consumable product is something that is disposable (razor blades, for example) or needs to be replaced after only one or two uses.

Bruce Johnson suffers from allergies. He also has a deviated septum, which constricts airflow through one nostril. His nose was always congested and his breathing problem made it difficult to sleep at night. Often at night he would awaken himself with his own snoring. He tried medications to solve the problem, but ended up addicted to nasal sprays. He tried putting padded paper clips, custom-formed wires, and shortened straws in his nostrils to keep his airway open, but those items either irritated his nasal passages or fell out while he was sleeping.

In 1988, it finally dawned on him one night as he was unable to sleep, that he might try something outside his nose—perhaps a device to go across the bridge of his nose to hold the airways open. It took him three years of fine-tuning to come up with a spring-loaded adhesive strip that would work to relieve his congestion. His sleep quality improved dramatically and soon he was ready to share his invention with the world.

Having taken the time to develop his product and knowing he now needed to test his product, he applied for a provisional patent and called an old friend who worked for CNS, Inc.—a company that focused on sleep-diagnostic equipment. Less than a year later (1992), Bruce signed a contract with CNS, Inc. giving them exclusive worldwide license for his invention of what is known as Breathe Right nasal strips. CNS, Inc. reported $22.9 million in the second quarter of its 2007 fiscal year—an increase over the $18.4 million in sales the prior year. Now that's a perfect idea!

Think of something for the masses—something that millions of people need that is not already on the market—and create a consumable product that meets the need.

Secret #6: Never Give Up!

Sometimes the market fails to recognize its geniuses. If people aren't seeing the benefit of your product, it could be so revolutionary that the market has to "wise up" at bit before your product can be accepted. Just think of all the inventions we have now that we can't do without. When microwave ovens first came out, people were wary of them because they were new and unproven. Surely, those radiation waves were dangerous! And, which one of us doesn't have a cell phone these days? If someone didn't push the edge and try to launch new products to the market, we would still be listening to 78s on a Victrola.

What did we do before the first word processing machine hit the market with its backspace feature that erased typographical errors and let us type in a correction? We used a lot of Wite-Out or Liquid Paper, that's what. But just imagine the frustration typists endured before there was correction fluid. If you made a mistake, you would have to retype the entire paper!

Bette Nesmith Graham was especially frustrated because she was not a good typist. Yet, she was a divorcee and had to work to support herself and her son, Michael Nesmith, who would later become a member of the Monkees. Out of desperation, Bette turned her kitchen and garage into a laboratory where she began mixing things together trying to come up with some type of substance she could "paint" over her typos, then type over it once it dried. She presented her magic formula of vinegar, water, and tempera paint to IBM in the mid-1950s as a product called "Mistake Out." IBM turned her away, but she was determined to get her product into the hands of other despairing typists. She changed the name of her product to Liquid Paper and sold it out of her garage for the next seventeen years. She was actually making a profit in 1979, when the Gillette Corporation offered her a deal she was too wise to refuse. They bought Liquid Paper for $47.5 million and additionally agreed to pay her royalties.

Secret #7: Question Things in Nature That Aggravate You

Simple annoyances may inspire your next invention. An everyday person can use his ingenuity to solve his own problem and help create a product that will benefit millions of people.

In 1941, a Swiss amateur mountaineer named George de Mestral took his dog on a nature hike. The duo returned home covered with burrs. Curious as to why those annoying little seed-sacs stuck so well to his wool trousers and his dog's fur, Mestral looked at the plant seed under a microscope. The plant seed had saw small hooks that reminded him of a crochet hook. These hooks had attached to the tiny loops in the fabric of his trousers. Through trial and error and in collaboration with a textile weaver, this persistent inventor designed a two-sided fastener; one side with stiff hooks like the burrs and the other side with soft loops like the fabric. Fourteen years later he had created a successful prototype that was to rival the zipper and all other types of fabric fasteners. In 1955, Mestral patented his hook and loop connection under the name Velcro—a combination of the words velour and crochet—and formed Velcro Industries to manufacture his product. Today, Velcro is a multibillion-dollar industry. Not bad for an invention based on Mother Nature.

Thomas Edison's famous quote, "innovation is 1 percent inspiration and 99 percent perspiration" is an excellent description of the invention process. The 1 percent inspiration is the creation of a remarkable idea. This can occur in a single second of inspiration—an idea hits you out of the blue, and Eureka! Suddenly you have a million-dollar idea in your head. Innovation is 99 percent perspiration refers to the time-consuming hard work of turning your idea into a viable product, and then taking the product to market. As in Mestral's case, the process can take many months, even years, before you see your product on the market.

When you create the right product, with the right name, for the right audience, at the right cost, during the right trend, and the product delivers more than it promises, you have a perfect product that may create its own new category and generate millions of dollars. If you believe your product is noticeably different in its unique shape, design, or function from similar products on the market, then it is time to proceed to the next step in the process: designing the prototype. But, first let me offer some advice so you don't become a scam victim along the way to inventing yourself rich.

Chapter Two

Steps to Inventing Your Riches

"A new idea is first condemned as ridiculous, then dismissed as trivial until finally it becomes what everybody knows."
—William James

The secret to your riches lies in your ability to manage the invention process. You want to tap into the marketplace with a new idea without running into legal snags or being ripped off along the way. This means managing every step of your invention, even if you hire someone else to do the parts you are unable to complete. Management is a skill that must be acquired; no one is born with it. Staying in charge of your invention will take time, money, and effort on your part. You will need to interview and research, and discipline yourself to spend your resources wisely when it comes to finding the best professional help, investor, designer, manufacturer, and licensee to work with. Do as much of the work as you can for yourself.

Read books, talk to other inventors, get advice, make your own prototype (if possible), and ask a lot of questions.

If you have a remarkable idea, you may as well expect that someone will try to copy it and beat you to the market regardless of whether or not you have a patent. Remember the story about the patriotic car ribbons? If you do encounter legal problems, you will need to decide whether or not to sue and try to recover your losses. It may not be worth the time and expense of hiring an attorney to fight a case that you have little chance of winning. Instead, focus on getting your invention to the marketplace where it can begin to create riches for you.

Secret #8: Focus on Selling, Not Suing (My Story)

While working as a personal trainer, I was in my office one day thinking about how to help a client of mine who was recovering from a car accident. She was experiencing terrible neck pain when attempting to do sit-ups. As most people do when performing abdominal crunches, she was putting her hands behind her head and pulling up. This was putting strain on her neck. I wanted to design something that would isolate the abdominal muscles and help her and other users perform a perfectly formed abdominal crunch without putting their hands behind their head. I found a paper clip in my desk drawer and began twisting and bending it until I created a small rocking chair design that I believed would work.

I went to the hardware store and bought some electrical conduit, a pipe bending tool, a roll of duct tape, and some foam. I sawed, bent, taped, and formed the materials until I had a full-sized, functioning prototype of the idea I had modeled with the paper clip. I hid the invention under my desk and confidentially asked some select clients, one by one, to try it out and give me feedback. Everyone loved it, and I felt sure it would be successful, especially since my patent research effort showed there was nothing like it was on the market. Next, I filed for a patent, located a tube bender and an upholsterer. Together we made fifty Ab Trainer products, which could sell for $100 each.

With prototype in hand, I attended a fitness show in October 1994. My small booth attracted the largest crowds at the show. Everyone was asking questions, and I received several licensing offers from the biggest players in the industry. One major fitness company offered me $ $2,500 and a 5 percent royalty. I was certain my product was going to be a big success and I didn't want to sell out before I had a chance to see for myself how well it would do. When I wouldn't sign the agreement, the company threatened to copy my design. They knew my patent had not yet been issued. Nowadays there is a provisional patent which allows you to put "patent pending" on your product. It gives some protection against patent infringement, but you will still need to apply for the full patent within one year of filing for a provisional patent.

I found a strategic partner and made an agreement with a major New York-based fitness company, which offered me $175,000 and a 50 percent share of the profits. They agreed to invest $300,000 for manufacturing and infomercial marketing. Through the fitness company, I sold hundreds of units, and within a year the Ab Trainer was selling to health clubs nationwide. Then, the bad news hit. A "knock-off" company had copied my design and was selling the product on QVC under the name Ab Rocker. Then, there was more bad news. The board of directors for the fitness company fired its president and cancelled the deal with me because they didn't think the product would sell on TV.

I threatened to sue the knock-off company, but my patent still had not yet been issued. The company offered to pay me royalties of less than a dollar per unit, which I refused. I focused on settling my breach-of-contract claim with the fitness company. They offered me an interest-free loan for $300,000 and I took it. I hired Stilson and Stilson, the company that had marketed the Health Rider product, to develop an infomercial for me. While the show was still in production, I got a phone call from my mom. She called to tell me she was watching the filming of an infomercial for my product in a local New Jersey mall. I didn't have the heart to tell her that it wasn't my product—it was another knock-off. I had no idea so many companies would rip off my invention.

I needed help in competing with the knock-offs that were appearing everywhere. I made a 50-50 deal with a California marketing company that paid me $200,000 in advance for the rights to my design. My Ab Trainer was a huge success as an infomercial item, with sales exceeding $100 million. After my patent was issued in 1996, there were twenty-seven other companies knocking off my product. Several were running infomercials and selling their product to retail stores. I sued and spent over $11 million to either shut down, win, or successfully settle with every one of the knock-off companies. Because my invention was so unique, we were able to obtain very broad patent coverage. Today, if it's called an Ab Roller, Ab Sculpter, Ab Toner, Ab Trainer, Ab Revolutionizer, ProTrainer, or any "Ab"-type product that rocks or rolls, you can bet my patent covers it.

While we are on the subject of patent infringement, let me remind you that suing the infringer may not be such a wise idea. The attorney that helped you get the patent is probably not a litigator and has never set foot in a courtroom. And, if your attorney is able and willing to represent your case, you should expect to pay a huge retainer (money paid up front) since contingency representation (pay only if you win) is generally not an option. But, the patent protects my rights, you may say. A patent gives you the right to sue for infringement, but it does not guarantee you will win the case, since the government will not fight for you. There are lawyers who chase ambulances, but patent attorneys do not chase patent infringers. Since there is no patent police, and it could cost you more than $1 million to fight for one patent through the court system. Unless you have a billion-dollar idea you might want to out-market rather than sue.

Secret #9: Protecting Yourself Without a Patent

You cannot stop an infringer until you have a printed patent in hand and an order from the judge, but there are other ways to protect your idea while you develop it into a product, without spending a lot of money for a patent. For example, when you share your idea with anyone, you should only do so if the other person or company has signed a non-disclosure agreement (NDA). This is a statement that binds them to respect the confidentiality of your work; a sample NDA is printed at the end of this book and also on my Web site, www.inventyourselfrich.com. The NDA will ensure that you will not lose your right to patent your invention later, and also ensure that no one can steal your idea without breaking the law. You would be surprised at how many people do not know about NDAs. For example, a friend of a friend came to see me one day and told me he had an idea that would make him millions. He knows that I'm a successful inventor and that I could help him.

"What's your idea?" I asked him.

"I can't tell you until I patent the idea," he replied.

"Don't be ridiculous, I will sign an NDA that will protect you better than a patent. Tell me about your idea, and if it makes sense, and if the idea can be profitably marketed, then maybe you should patent it."

My friend was stunned. How could he possibly share an idea with me without first obtaining a patent? What if I copied it? What if I took it and gave it to someone else? How would he protect himself without a patent? Obviously, he didn't know about NDAs. If a person or company refuses to sign your NDA before you share your idea, then you should not share your idea with him. Once they sign your NDA, you can share your idea, and begin the process of determining whether your idea is valuable and whether it can be marketed profitably.

The U.S. patent system rewards and protects the first person who invents a new product. What that means is if two people are working on the same idea, at approximately the same time, the patent will be awarded to the person who can prove that he or she conceived the idea and invented the product first. As soon as you have a remarkable idea, you should begin keeping an inventor's logbook which shows the chronological development of your idea. **The logbook that shows the earlier start date will win, even if the other person filed the patent first!**

When two people are contesting and claiming the same invention or idea, the person who has the most detailed, dated entries and sketches will win the legal battle nearly every time. Therefore, you should include drawings of your product in the journal, and sign and date each entry. The pages of your notebook should be witnessed weekly by two people you trust, but are not relatives, and who understand the significance of what they see and read.

Do not be concerned that your logbook will document the paths that turned out to be dead ends as well as those which were fruitful; no one expects you to have an idea and then turn it into reality without any alterations along the way. The inventive process can be long, and the role of the logbook is to document the path that you took. In addition, your witnesses should sign a non-disclosure agreement (NDA)—and allows you to discuss your invention with some protection. Also, you may want to consider getting the pages notarized by a notary public.

The type of logbook you choose is important and must meet certain criteria. The back section of this book is a logbook you can use for one of your inventions or you can obtain a stitched or hardbound notebook (like black-and-white covered composition books) that has numbered pages so there can be no mistaking when pages are missing or torn out. An accountant's ledger is a good choice. Avoid spiral notepads, glued tablets and three-ring binders with loose sheets as they will not stand up in court. A book with graph paper or gridlines works well for drawing to scale.

Use an ink pen rather than pencil to write in your logbook to make sure your entries aren't erased by a third party. By the same token, you should draw a large "X" through any blank pages to keep anyone from adding anything. If you are entering a drawing, you may want to first practice on a separate sheet of paper and then copy the drawing into the logbook. You don't want to cross through any text or drawing or rip out any pages. If you make a change to an original entry that has been witnessed already, the original witness should sign and date the change. This book contains sample logbook pages, as does my Web site, www.inventyourselfrich.com.

Keep all paperwork, invoices, and receipts for materials you used to create prototypes, and staple them to the pages of your journal. Log in your mileage and dates of visits to patent attorneys, manufacturers, or other people you deal with along the way and get their signatures in your log as well. Include the names of anyone you share your idea with, and staple a copy of the non-disclosure agreement to a page in the journal. List Web sites and books you read that are related to your project research. Record any test results, diagrams, marketing ideas, etc. Each entry should be written legibly and be self-explanatory so that anyone can understand it. Be sure to date (include the time) and sign each entry. Show a continuous and persistent effort along the timeline in order to meet the patent office's "due-diligence in completion" requirement. A lag time of only a few months could endanger your ability to obtain a patent. To avoid confusion and show a chronological record of all things related to your invention, keep a separate logbook for each of your inventions. Because the Inventor's Logbook is such an important tool for recording the process of your invention, I am including one as part of this book.

Keeping a logbook will help you fight any competitors who try to steal or copy your invention. I am reminded of the story about a small inventor who created the idea of a "two-in-one" stylus pen to be used for the popular Palm Pilot devices. The inventor kept detailed sketches; every minor change was dated, creating a complete picture of the product's development. Whenever the inventor showed his sketches to anyone, he had them initial and date the page. The inventor believed that 3Com Corporation, the maker of the Palm Pilot, would be interested in his invention, and indeed they were. During months of meetings, where the inventor shared his idea and all of his drawings, he wisely made the 3Com executives sign and date each sketch in his pad. 3Com decided they would make their own two-in-one pen without the inventor. After initiating a costly lawsuit against 3Com, the inventor was able to settle the case without going to trial. The logbook proved to be powerful evidence against 3Com and they decided not to fight the case.

The bottom line on protecting your invention is to make sure you notate your invention in a logbook from the beginning, make sure you use an NDA before disclosing your invention to anyone, and make sure you file a provisional patent application before you disclose it publicly. A provisional patent application is not examined and can never issue as a patent. It serves as a patent application and gives you one year to file for a full patent.

Design the Prototype

Once you confirm that your idea does not infringe another patent, and that your idea is unique or different than other products on the market, the next step in the process involves turning your idea into a working product. This means you need to create a prototype of your invention—a real, working embodiment of your idea such as the coffee jacket I described in the beginning of this book. Many new inventors think they can skip this step, or just license their idea to a company. But this never happens. Companies don't care about ideas—they care about products. So it is crucial that you create a prototype of your invention as a first step toward determining if you have a viable product that actually works.

While a prototype is a real working model of your idea, it does not need to be perfect. This is not the final version that will go to market; the prototype is basically a homemade model of your idea. You can create a prototype yourself or hire someone who can make the prototype for you. The cost of a prototype will depend on the time and materials needed. If you cannot make a prototype yourself, there are professionals who can help you. You may go to my Web site for this information.

Exactly what degree of perfection should you try to attain in creating your prototype? It depends upon the type of product it is. You can start with materials you have on hand, but if the finished product is going to be made of molded plastic you will probably need an industrial designer who can create a prototype that looks and works like the eventual product. The appearance, functionality, and packaging are important for your "show and tell" demonstration. If you are planning to show the prototype to investors you need more than cardboard and duct tape. The prototyping process is not as scary as it sounds. You may be able to do it yourself like I did with my Ab Roller invention. After I came up with the idea for the Ab Roller, I borrowed a pipe bender from an electrician and went to Home Depot and purchased a piece of electrical conduit (metal tubing), then went home and built the first "rough" prototype of my invention.

Once you have a rough working prototype of your invention, you can begin showing it to other professionals for input. But only share your prototype with reputable professionals after they have signed a non-disclosure agreement. These professionals, such as design engineers and mechanical fabricators, should provide feedback on how best to manufacture and design a real working product based on your invention.

Not all inventive brainstorms will need a product prototype; some are intellectual property, special programs, or financial packages rather than a tangible product you can hold in your hand. These non-tangible products are protected by copyrights and trademarks.

Funding Your Invention

Up to this point in the invention process you may have spent about $1,000 to $1,500 for a patent search and materials for your prototype; that is IF your prototype is something you can create yourself. From this point on, you will need a lot more money to have a professional draw the blueprint, manufacture a sample of the product, and hire a patent attorney to draft the patent application. If you don't have money on hand you will need to find another way to fund your invention. No matter how good your credit rating or how well they know you, the SBA and most banks will not loan money for prototypes, molds, packaging, attorneys' fees, and marketing efforts. A lack of funds will hinder your chances of success, so if you don't have a family member who will lend you the money, you will have to get creative to find ways to move forward. Even if you have family members willing to lend you the money, make sure they can afford to lose it. No invention is a sure bet.

T.J. DeFlavis took a personal loan for $3,000 that his parents matched in order to develop a prototype of the Hook Up. His invention is a hook with a strap that closes itself around extension cords, tools, camping gear, boating accessories, and garden equipment to hang it out of the way in the garage or on a work site. T.J. used the loan and his parent's money to manufacture 1,000 units and he pitched it to tool centers, home and garden shops, and rental centers. He was doing his research throughout the project. He read an article in a magazine that connected him with Angel-

Guard, where he made a licensing deal that will make the Hook Up a common feature in garages, campers, job sites, and boats everywhere.

Mark Davis had an "eggcellent" idea for an egg-shaped hand exerciser that would strengthen all the fingers. He needed $26,000 to create his prototype and pay for legal fees. He decided to start his own company in which to market his egg. He sold shares at $2,000 each to interested partners. He raised $18,000 in three weeks! Then, he called in favors and bartered with friends and family who could help with packaging and promotion. Within thirteen months he had a market-ready product that he took to the National Sporting Goods Association Trade Show in Chicago. Sales were slow there, so he took out a few ads locally to let people know about his egg. He was featured in several sports magazines and sold nearly 90,000 Eggsercisers, but he was still barely showing a profit. The huge hit came when he obtained national exposure through a story written by John Peirson in the *Wall Street Journal.* Peirson interviewed a doctor and a coach and mentioned Brookstone—a specialty chain store that was selling a competing product. People started calling Brookstone specifically asking for the Eggserciser. Brookstone placed an order for 2,500 eggs and sold out within a week. The Eggserciser sold more than 700,000 eggs that year and they are still selling ten years later.

It might be difficult to find people who are willing to invest in the future of an unproven product. People do not want to lose money on a fancy idea no matter how exuberant you are. They will invest in facts, not hype, so you will need to remove all doubt. This is where a marketing plan is a good idea. In the plan, you must provide evidence to investors and vendors that the product will perform as you say it will. You will also need to outline the path you plan to take in bringing the product to market and show a cost estimate of what it will take to turn your concept into a financially solid investment. Include expected profit margins and a time table showing process of production. Anticipate the questions investors might ask, and then answer them in your marketing plan. Once you are in front of a potential investor, demonstrate how the product works and ask the investor to "play" with your toy or conduct a trial use of your product. Paint a picture of how much better the world will be when it has accepted your invention. In other words, sell your invention idea before it becomes a product.

Select companies in your field that might have an interest in backing your invention. Start contacting them by telling them you are a product developer (don't say that you are an inventor) and ask to speak to the director/vice president of marketing. Find out if they have a standard form for submitting ideas and ask for the name of the decision maker—usually the company's president or vice president. Submit your proposal via their guidelines or ask for an appointment to make your presentation. Have your marketing plan ready to present to large corporations, bankers, brokers, businesspersons, and anyone with a wad of cash in their pocket who will listen. And, like any good salesperson, don't take "no" for an answer.

If your invention is going to need more money than you think you can generate from investors, you may want to subsidize your funds by working a second job or applying for a grant. I have listed several resources in Appendix B in the back of this book. You can use the money you raised from family members, your second job, or from selling company stock to early investors to create a professional quality prototype before consulting with an investor.

If you are inventing a medical or high-tech device, you might want to work with a venture capitalist. This type of investor is more likely to work with your start-up company than with your product, so you will need to show that you have key players in place, each having a record of solid accomplishment. These "angels" will usually invest $2-$5 million if your company is secure and your product is a well-developed, patent-protected invention in their company's area of interest. You will have to provide proof that you have a market and can reach it. You must be able to assure them that your company and your product will create extremely high return on their investment. I don't recommend that a novice inventor try to deal with a major corporation alone. It's a huge undertaking, so let your attorney work with you if you decide to go this route.

Production Drawings

Next you need to convert your prototype into professional engineering drawings. These drawings are important to a manufacturer so your product can be made according to precise mechanical specifications. Professional engineers who perform these services will also help to tweak the design of the product and improve its functioning. Good professional design feedback is invaluable at this stage in helping to turn your idea into an innovative product. Again, be sure to use a non-disclosure agreement.

Marc Schneider used Adobe Illustrator to create his mechanical drawings when he invented The Stinger Stylus. His adjustable-size ring has a plastic "pencil" tip that is placed on the first joint of the index finger. It is more easily used than the stick stylus that comes with hand-held computers such as the Palm Pilot. If you are adept at working with mechanical drawing software, you could create the plans and have an engineer look over them before you take them to the manufacturer. A design engineer will be able to create cross-sections and computer-animated drawings that allow you to "see inside" the working of the product. You will find resources on my Web site: www.inventyourselfrich.com for reputable professionals to help you with design and drawings.

Create the Real Thing

Once you have professionally engineered drawings, you can create an actual working product—not just a prototype—but a real sample of the finished product. Rather than spending lots of money on thousands of units from a manufacturer, you can instead use your money more efficiently by creating a "pre-production" version of the product. There are companies who specialize in this type of manufacturing, and they will run a small quantity of units for you. The price per item will be much more expensive than if you manufactured thousands of units, but why spend tens of thousands of dollars on tons of inventory if you do not know the product will sell? Plus, at this stage, you may continue to make design changes to the product, and you do not want tens of thousands of units of your unrevised product sitting in your garage.

As manager of a 24-hour Walgreen's store, Jeff Kempher had seen several employees cut badly enough using a traditional exposed-blade box cutter that they had to go to the hospital for stitches. Kempler came up with an idea that would solve the problem. He tried as many as forty styles in a variety of materials, but he didn't know what steps to take next. The finished product has a plastic handle with a mushroom-like head that wraps over the sides of the device where the blade is tucked inside. The plastic edge is exposed to the user and the package, preventing product damage and user injury. He mentioned his idea to a sales rep who called on his Walgreen's store. Orville Crain, who owned a wholesale distribution business, linked Kempher with Matt Jacobs, who owns a manufacturing company. Crain, Kempher, and Jacobs formed a company and together rolled Klever Kutters off the production line in the U.S. for $2.99 each. Soon the word was out and they had grocery chains, warehouses, and home stores calling to purchase the tool.

Sourcing Your Product

Now you can take your pre-production sample to a large-scale manufacturer to determine the best way to mass-produce your product for maximum value. This step may also require further design changes to the product, depending on how the product can best be manufactured. This step is called "sourcing" the product. Again, be sure to use a non-disclosure agreement when showing your design to a manufacturer.

There is no need to actually order product from the manufacturer. At this point, you just need a sense of how much your product will cost to manufacture in quantity. When you have this cost in hand, you will then be able to know how much you can charge for the product at retail.

John Higgins came up with a great idea for a pill case that can be carried in a wallet. The Re-Pillable card that holds five aspirin was manufactured by ProtoPart, Inc. in Hudson, New Hampshire—a company Higgins found in the Yellow Pages under "plastics" (www.protopart.com). They worked with him to get it produced at a price point where he could make a profit. He took an actual product to *Men's Health* magazine, the source of his inspiration, and the magazine featured an article on the Re-Pillable.

Secret #10: The Price Must Be Right

To quote Thomas Edison: "There is a wide difference between completing an invention and putting the manufactured article on the market." You may have a great working product, but the next consideration is whether it is a viable or marketable product that can be sold to the consumer for a large enough profit to make it worthwhile. By marketable I mean whether all the parties involved in the commercialization process of your invention, including yourself, can make money by marketing your invention. If all the parties in the process cannot make money, then it will be an uphill battle to get your product to market.

The single most important fact to understand in making this determination is the ratio of retail price to manufactured cost. The ideal ratio is 5:1. This is a very important fact, and I will take some time here to explain what it means.

A 5:1 ratio of retail price to cost means that a product selling for $50 to a consumer in a retail store must be manufactured for $10. I can tell you that most people do not understand this concept. Many inventors believe that if they can manufacture their invention for $25 it could then be profitably sold at retail for $50. This is just not true. The 5:1 ratio is an ironclad rule of retail. The only exception is if you are selling over the Internet, direct to the consumer, then your ratio can be lower and still be profitable. If the ratio is bigger than 5:1, that's great—the product will be even more profitable. My Ab Roller had a selling ratio of over 8:1, which made it one of the most profitable products of all time! If the ratio is lower than 5:1, the product may still be marketable, but it will be more difficult to generate a profit for everyone, especially investors.

With this 5:1 ratio rule in mind, you can see that many inventions simply cannot go to market and thus are not marketable. For example, jump ropes sell at retail for anywhere from $5 to $20. If you create a truly innovative jump rope that costs $10 to manufacture, then you will never be able to take it to market. Unless people would be willing to pay at least $50 for your new jump rope, there is no way this product can be profitable. Even if it could sell at retail for a premium price of $30, there would still be a shortfall of only a 3:1 ratio.

Why is the 5:1 ratio an ironclad rule of retail? It's because there are many costs involved in taking a product from the factory floor to the consumer's hands. There are costs that need to be paid as a product moves from drawing board to factory floor to warehouse to sales representatives to retailer and finally to customer. Each link in the supply chain costs money. Money for raw materials, labor, equipment, services, transportation, warehouse space, manufacturing space, office space, shelf space, packaging, advertising, customer service, telephones, inventor's royalty, legal and accounting fees, taxes, etc.

Many inventors are also surprised that the wholesale price paid by a retailer for your product is typically one-half of the ultimate retail price. In other words, if your jump rope sells for $20, then the retailer will pay you only $10 for the rope. The retailer needs to sell products at double the wholesale price in order to cover its costs and make a profit. So, if it costs $10 to make the rope, you will not make any money.

Table 1 shows a rough breakdown on where the money might go for a prototypical new product retailing for $20 at a mass merchant, utilizing the 5:1 ratio of retail price to cost.

Table 1.

MANUFACTURER'S COSTS:	
Product: labor, materials, in-bound freight, tooling, etc.	$4.00
Other Business Expenses: (engineering, marketing salaries, rent, royalty, etc.)	$3.50
Total Cost	$7.50
Wholesale Price	$10.00
Manufacturer's Net Profit	$2.50
RETAILER'S COSTS:	
Product Cost (Wholesale)	$10.00
Business Expenses (salaries, rent, marketing, etc.)	$5.00
Total Cost	$15.00
Retail Price	$20.00
Retailer's Net Profit	$5.00

To manufacturers and retailers, the product itself is really insignificant. They are in the business of building and selling boxes that contain profits. A new product is an opportunity to sell more boxes and reap additional profits. They don't care what is in the box as long as it turns a profit. Those boxes/products with the highest return are the ones that continue to get shelf space. Products that fail to pay a profit are evicted. Shelf space is too limited to waste on a product that doesn't make money.

Secret #11: Control Manufacturing Costs

One way to reach the 5:1 ratio is to keep the cost of production down. If you can work out a way to keep your manufacturing in the U.S. you will have less risk, quicker deliveries, and more control. Sometimes all this takes is a minor adjustment in the design, as in the case of Nukkles. There are over 10,000 differently shaped hand-held massagers in the world. How did Nukkles break through and become the number one massager? Through its unique design. Nukkles looks very cool, and its thin 4-dome design bends and flexes, acting like little shock absorbers. Most other massagers are rigid and hurt when pressed on bony areas of the body. Nukkles design also allowed anyone to give a perfect backrub without having tired hands.

I did not invent Nukkles; instead, I marketed it. A woman named Myra Per-Lee from California was the original inventor. She was selling Nukkles at a fair in her town when a friend of mine who was a personal trainer purchased a pair. My friend demonstrated them to me by rubbing my back and that was all it took. I quickly called Ms. Per-Lee and told her I wanted to produce an infomercial. We agreed on a royalty and away we went. I redesigned her package, invested $75,000 in a new mold that would make six Nukkles at a time, and then worked out a deal with the molder so he would mold, package, hot-stamp, and ship out my orders, all from his one location in

New Jersey. This deal allowed us to keep the product in the U.S. and not have to deal with long lead times from importing products from China. Remember, every time someone has to touch your product or the pieces to make it, your cost will go up. That's why the more you can get done under one roof, the less cost you will have in your product. Besides that, "Made in America" is a great marketing tool.

If you decide to re-invent the wheel, make sure it offers something the others don't for a price the competition can't beat. The bottom line is to make sure your invention can be marketed profitably before you attempt the long process of obtaining a patent and licensing your invention. Make sure your product is the type that can be sold for at least five times the manufacturing cost. You can perform this analysis informally by comparing your invention to similar products on the market to find the potential price consumers are willing to pay; and then estimating the manufacturing cost of your product to see if the 5:1 ratio applies. If your product passes this 5:1 ratio checkpoint, you are ready to begin patenting and licensing your product and taking it to the market.

Ready, Set, Go Get the Patent!

Once you feel comfortable that you have developed a product that can be sold profitably, then you should think about obtaining legal protection, such as a patent and trademark. Applying for a patent can be complicated due to the laws which govern it. Drafting a patent is the most difficult of all legal writing. Unless you have a legal background and understand how to draft a patent you would be best to hire a patent attorney to help you with the legal requirements once you have a prototype of your product. If you don't know what you are doing, you may end up filing a patent application that doesn't give you adequate protection for your invention or it may slow down the process. If you want to save a few bucks, you could write the patent application yourself, but it would be a good idea to have a patent attorney make sure the wording is correct. Remember that if a patent is not written correctly is will be useless, so do your homework and find an attorney who has the appropriate technical background for your invention. Plan to work only with a patent attorney or agent who is registered by the USPTO, and meets legal, technical, and ethical qualifications. To be registered with the USPTO the attorney or agent must have a degree in engineering or science in addition to passing the patent bar exam. Since the USPTO has no jurisdiction over patent attorneys or agents, and there could be unscrupulous ones in the system. When interviewing prospective attorneys ask each one how many patents they have filed in the past year (he or she should have filed 10 to 25). Also, be sure to get a price quote for the entire patenting process.

Facts About Patenting

Since its inception in 1790, the U.S. patent system has played a major role in the world's economic affairs. The system was set up to provide an incentive for creativity, to promote the advancement of commerce and industry, and to prevent monopolies in the United States. A patent gives an inventor exclusive rights to his or her writings and discoveries for a limited time. Today, a patent protects the inventor's product for twenty years, but it requires that the workings of their invention be disclosed to the public in order to specify what others must not make lest they infringe the patent of another. Patent holders may market their own product, or they may sell or lease the patent rights to another person. After twenty years the utility patent expires and anyone can make the product.

A patent is granted to the owner of the invention or the person the inventor assigns as owner of the invention. A patent will include: the name(s) of the inventor(s), a claim to the invention, a conceptual statement describing the invention, drawings of the invention, the name of the invention, the purpose and advantage of the invention, and an explanation of how to make and use the invention.

What Is Intellectual Property?

Due to the growth of the industrial revolution and technological advances, the rules regarding what defines a "new" product and what may be patented have changed much over the years. Today, there are four classes of protection for intellectual property: patent, trade secret, trademark, and copyright. Since we are not dealing with music, films, photography, or literary works, we will skip copyright, and since there are no federal laws protecting trade secrets (chemical compounds, formulas, machine patterns, technical know-how, customer lists, and software) we will let you research the requirements, which vary from state to state, for patenting trade secrets. We will review trademarks in the section on marketing. Our focus here is on the patent.

A patent is a legal right issued by the federal government and gives an inventor the right to exclude others from making, using, leasing, selling, offering to sell, or importing an invention in the United States. However, you may find it surprising to learn that a patent does not guarantee the right to make, use, or sell your invention.

What Can Be Patented?

A patent doesn't protect as much as you might think it should, and patents apply only to inventions that fall within legally defined categories. Anything that lies outside these categories cannot be patented. There are three legal classes of patents: utility, design, and plant patents. Since most people reading this book are not botanists, I will skip over plant patents. A utility patent is given for chemical compounds or formulas, new processes, devices, machines, and manufactured items; in other words, it covers just about anything that can be made. A utility patent is good for twenty years before it becomes public property. A design patent only covers the appearance of a manufactured item and is good for fourteen years.

There are also three statutory classes that your invention must meet in order to be patented: novelty, utility, and non-obvious. For it to classify as novelty, the invention must be new. If something is already advertised or on the market (even without a patent) it cannot be patented. Word of caution: Do NOT share your secrets or publicly describe your invention via email or Internet, printed material or at a conference or trade show without a signed NDA in place even if you have applied for a patent. Remember a patent only covers certain aspects of the invention. Selling or offering to sell your idea or product is considered public disclosure, which may cause your invention to disqualify for a patent.

For your invention to classify as a utility, it must be useful or perform a function that benefits society in some manner. You can't say the invention does something if it does not perform its intended purpose. For an invention to classify as non-obvious, is must be a "eureka" idea and offer some function or design that would make people say, "I should have thought of that!" The idea wasn't blatantly obvious.

It may take the USPTO one to three years to grant or deny your patent application. For this reason, I recommend you file a provisional patent, which is inexpensive compared to a full patent application. A provisional patent does not have the same legal requirements as an actual patent application, but it establishes a patent date and permits the term "patent pending" to be applied to your invention while you continue to make final improvements. You will have one year after filing the provisional application to file the actual application.

After you have an idea, you need to make a prototype, do market research, and then file for a provisional patent. Make sure you do not publicly disclose your idea during the first two steps. While it is important to get a provisional patent, you do not need to wait for the final patent to be approved before you share your idea if you have a signed NDA with each person you talk to about it.

Secret #12: License Your Product or Start a Business

After you have determined that your idea is marketable, and after you have turned your idea into a viable product, it is time to take it to market. This is probably the most challenging aspect of the invention business. When your product is ready to take to market, you are entering a whole new phase of work that requires sales and marketing expertise. Before you invest your time and money in your product you need to decide whether you want to license your product to a third party and make a royalty of 5 percent (or less), or if you want to start your own business and launch the product into the market yourself.

You may be ready to sell ownership of your patent and move on to your next invention. If so, you will contract with an individual or company to receive an agreed-upon payment and let go of any future royalties. I don't recommend this route because you never know what commercial potential a product may hold, and you may deprive yourself of a fortune. Just think of the offer NordicTrack made me for my Ab Roller. I would still be kicking myself if I had taken their original offer.

Try not to be too emotionally attached to your invention. Sometimes it is wisest to have a third party involved. If you decide to go that route, you need to do your homework and interview prospective licensees to find the best fit for your invention. Companies are primarily defined by a particular technology or market. When looking for a licensing partner try to match your product with a company that already has a base of customers who would buy your product and an established channel of distribution to the market.

In partnering with an invention licensee, I suggest you retain ownership of your patent and make a deal for royalty payments while giving that company or individual a license (permission) to make and sell your product. You want to make sure the agreement is fair for both the licensee and yourself. There are a few things you should make sure your contract states:

That your non-refundable advance against royalties is based on sales rather than profits

The percentage of royalties you are entitled to; for how long and when they will be paid

Legal right for you to see the licensee's sales records

That your name is included on the licensee's product liability insurance policy

How long the licensee has to bring your product to market

Who pays for legal fees to complete the patent or if there are infringement issues to battle

What is to be done with remaining inventory

What territories are you awarding (U.S., Canada, others)

Do you keep the rights to your invention? If not, is there a way to get them back later?

By having an agreement that allows you to retain ownership of your patent, you will be able to step in and take action if things aren't going as you would have hoped.

Don Chernoff was an engineer in Silicon Valley, California, who traveled a lot for his job. He shared the frustration of many frequent flyers: lost luggage, carry-on bags that would not fit into the overhead compartment, and having wrinkled clothes when he reached his destination. There simply had to be a better way to pack luggage. He concluded that if he rolled his suits rather than folding them, they would fit into a smaller suitcase and wouldn't wrinkle as much. Chernoff made his prototype from 8-inch PVC pipe with a garment bag wrapped around it. He applied for a patent on Skyroll, a garment bag that holds suits, shirts, or dresses and wraps around a hollow lightweight cylinder that holds toiletries, shoes, and small items of clothing.

The luggage industry was transformed when SkyRoll hit the market. However, Don Chernoff didn't want to go into the luggage business. Instead, he wanted to license his product, earn royalty payments, and let someone else handle the details of running a business. Chernoff hooked up with QVC to market his product on TV. (To learn about how to get your product marketed on QVC, see www.vendor.studiopark.com/howto.asp.) QVC referred him to a company in Florida to help him manufacture the product. After a year of designing a finished product, 8,000 units were made in Thailand. QVC sold 7,000 units in spite of an air time of 3 to 5 a.m., when few people watch TV. QVC wasn't satisfied with the sales and instead of airing the product at a different time, they dropped it altogether. When QVC rejected the product, the licensee lost all interest, so Chernoff took back his rights to the royalties and started selling his leftover inventory on his Web site www.skyroll.com. When he sold out of product and needed to order more, he found that working with people overseas was difficult. He finally received notice that the factory manager had quit and the company in Thailand was too busy to handle his business.

Chernoff started over again and found an independent sales rep who put him in touch with Don Godshaw, President of Travelon, a Chicago-based specialty travel product company. Chernoff licensed his product to Travelon, which redesigned the product and improved its structure, making it more durable and a little more expensive. This allowed him to tap into a more discriminating market of customers who fly frequently. Travelon's first order was 10,000 units, again produced overseas. Chernoff stayed in control of his product the entire way to the market while subcontracting some of the tasks he didn't want to tend to. Even though he spent a great deal of time and energy on his invention, he only spent $10,000 for his "perfect" idea. He easily made a profit on his investment as soon as the product hit the market the first time. Perseverance is one of the keys to a successful invention. Make sure you keep your goal in mind and push toward the finish line.

If you are an ambitious entrepreneur, you may want to start your own company, rather than licensing your product to a third party. This means you will have to fund, manufacture, warehouse, transport, sell, and distribute your product on your own. Before you attempt to start your own company, I advise you to contact the U.S. Small Business Administration (SBA) and create a business plan. You may also want to read legendary entrepreneur John Osher's 17 mistakes to avoid in starting a new business. Among the mistakes John Osher listed in a February 2004 interview with *Entrepreneur* magazine's Mark Henricks were these pitfalls:

Failing to spend enough time researching the business idea to see if it's viable

Miscalculating market size, timing, ease of entry, and potential market share

Underestimating financial requirements and timing

Hiring too many people and spending too much on offices and facilities

Lacking a contingency plan for a shortfall in expectations

Bringing in unnecessary partners

Hiring for convenience rather than skill

requirements

Focusing too much on sales volume and company size rather than profit

John Osher invented himself rich by starting and selling companies as a way to sell the products he created: Rainbow Toy Bar was a gym for infants that he sold to Gerber. He continued to work for Gerber as a vice-president until he started his company, which he later sold to Gerber. Next he started CAP (Child at Play) Toys, which nearly closed in the first year of operation when "a blooming doll in a flowerpot" didn't sell. CAP's second toy was an Arcade Basketball hoop that hung from the back of a door and kept score. That product's success was followed by the Stretch Armstrong doll and the battery-operated Spin Pop that rotated a lollipop on a stick. After falling upon hard times and going through a divorce, Osher sold CAP to Hasbro and became their "VP of Nothing" for about a year. Then, it was back to inventing. With new partners, he formed another company. Dr. John's Products launched a battery-powered toothbrush (designed similarly to the Spin Pop) in Meijer's—a chain of Midwestern department stores. The SpinBrush caught on quickly, and soon Osher and his partners were working with Procter & Gamble through a deal they struck to make sure the product was successful. Through their effort, SpinBrush sales have reached $300 million per year.

And, it shouldn't be surprising that a toothbrush would find such success. When 1,442 people participated in a survey by the Lemelson-MIT program, they were asked, "Which of the following could you absolutely not live without: your car, your computer, your cell phone, your toothbrush, or your microwave?" Forty percent of the respondents said they could not live without their toothbrush.

From inventor to business owner, business owner to inventor, again and again, John Osher's career has rotated like his SpinBrush, yet he is a successful entrepreneur in both ventures.

On the Organization Concepts Web site, Peter Russo, director of Boston University's Entrepreneurial Management Institute, provides additional tips to get a small business on the right track including:

Know and constantly review your goals for the venture

When you need help, recruit and hire the best people

Be honest with yourself (one of Jack Welch's mantras)

"America's # 1 Idea Guru," Doug Hall, also explains that for a small business to succeed it has to understand the "Three Laws of Marketing Physics" which apply to any product or service offered for sale. According to Hall, these laws are:

Overt Benefit—What is the benefit your service or product provides and can your customer identify that benefit immediately? Your idea must be clear, direct and focused. Your customer should instantly realize the benefits they are going to receive for investing their time, energy, and money. If a customer decides not to buy your product, make sure it is because they did not want it, not that they did not understand the benefit of your product.

Reason to Believe—Why should your customers come to you for your product? Your idea must deliver credibility to your customer by telling the truth about the benefits they are going to receive. Under promise and over deliver what your customer expects.

Dramatic Difference—What sets you apart from the competition. Your idea must be unique, different, and so much better than what the customer is currently using that they want to change. Your idea must create such a memorable experience that your customer will want to tell others about it.

All of these concepts for the successful entrepreneur also apply to inventing.

To Partner or Not to Partner

If you are like me you are always coming up with new ideas faster than you can finish the first one. Having a partner can help you stay on target. However, having a business partner presents a new set of opportunities and problems. They can help you invent yourself rich, or they can invent themselves rich instead and leave you without a penny.

Here are a few tips to consider before deciding to partner with another person to get your product to market.

Rules for Partnering

1. Never give up control of your product unless you are licensing it to a big company. Even then you should have minimum sales requirements.
2. Admit your weaknesses and delegate those responsibilities to others who are strong in that area. Only partner with individuals that add value where you are weak.
3. Always maintain 51 percent ownership in any of your deals.
4. Brainstorm together but you should always be the one who makes the decision.
5. Have a very clear exit plan if things go wrong.
6. Beware of having an attorney as a partner. They can easily take advantage of you.

7. Define exactly what the partner's role is and delegate those responsibilities.

I broke several of these rules for partnering with my Ab Roller invention because I didn't have any Golden Rules back then (see the Conclusion for a list of my Golden Rules for Riches). I learned from my experience and share this advice with you, so make sure you apply it.

Branding and Packaging

In order to stand out on a crowded store shelf—and even to get ON to a store shelf—your product will need a strong branding and packaging strategy. There are a lot of products competing for the consumer's attention. And shelf space in retail stores is limited because retail stores only want to carry products that sell, sell, sell! So, you need to create a memorable name and logo for your product, and you need to design innovative, catchy packaging with splashy graphics.

Apple computer is probably the best example of innovative design when it comes to packaging. Every detail is perfectly planned to create a memorable experience opening the package, now that is pure genius. I can't wait to see the iPhone package—you can bet it will be something you will never forget.

The basic function of packaging is to protect and contain a product, but in our competitive society we use size, shape, and color to create brand recognition through packaging. This helps differentiate the product from its competitors and gives it a unique selling point that makes customers want to buy it. The actual product is inside the packaging, but you have to attract people to try a product through packaging and advertising. Even if the product itself is not good, the packaging gives the impression that it is. I have actually bought products whose package design was better than the product inside! Unless you are a marketing expert, you will need to hire a designer and a packaging expert to help you with the artwork and product packaging.

Secret #13: Packaging Makes All the Difference

When it comes to packaging and marketing your product, remember that a load of rocks can make a ton of money! For example, Lava Ice is a natural soapstone selectively quarried from the Himalayan mountains. Lava Ice can be heated or chilled and will maintain a constant temperature for extended periods of time, making it perfect for therapeutic use. Each stone is hand-carved and polished smooth, with a body-contouring design on one side and a raised acupressure dome on the other side.

I've sold 5,000 rocks that retailed for $19.95 or more. They cost me $0.67 each. How can a rock bring so much money? It's the packaging and marketing that sells this product. We could have put it in a plastic bag and marketed it at Dollar General and still made a profit, but taking the packaging up a notch (the stone comes in a cloth bag in an attractive wood box) made it appeal to health-conscious individuals in a higher income bracket. Additionally, our marketing text was very compelling:

> LAVA ICE can be used like a hot water bottle OR an ice pack without the hassle or the mess! Unlike ice cubes and ice packs, LAVA ICE never melts! You can use LAVA ICE the way you would use an ice pack or heating pad, simply by placing it on areas of the body that are inflamed, in pain, or in need of cooling off. LAVA ICE is great for headaches, swollen/puffy eyes, muscle and joint soreness, cold hands or feet and arthritis. Simply heat or chill the stone according to the instructions, put it in the cotton bag provided, and apply it to the area where you need relief.

The combination of attractive packaging and superb marketing created a cost to manufacture ratio of 20:1.

Brand Name and Logo

Your invention is your baby, and during the "gestation period" you will undoubtedly be thinking of names for your brainchild. The brand name and image of your product are an important part of the branding process and should be well thought out before making a final decision. When naming your product think of it as a household term—something everyone can pronounce when they see it spelled. Large manufacturers know the importance of having a brand name. Kraft, Nike, and Microsoft are all household names which we associate with quality. Most people are likely to buy a name brand product when they go shopping rather than one that is untried or unknown—even if it's a store's own brand. Therefore, it is important for your brand name and logo to be prominent on the packaging.

Secret #14: The Perfect Name Is Almost a Lie

Some products carry a name that alerts the consumer instantly to know that there is more to the name than the product can possibly deliver. These name games trigger curiosity and instinctively make the potential buyer want to know more about the product. There is no way a person can get perfect abs in six seconds, so how can an inventor or manufacturer advertise his or her product as the 6 Second Abs? Because the apparatus is used to shape the tummy by working the abs for only six seconds at a time. The name represents the truth, but you have to use your imagination or dig deeper questions to discover the whole truth. And that is exactly what the inventor wants you to do. The same is true for the 3-Hour Diet at Home, which is a program that offers three different meals and two snacks delivered to your door. These meals should be eaten every three hours, thus validation of the program's name: The 3-Hour Diet. Is it logical to think that "The Ab Lounge" might be a place to lie around doing nothing while creating firm and toned abs? No, but seeing a photo of the product soon lets you know it is a lounge chair that is used to properly perform abdominal crunches. The secret is to give your product a name that describes the product without being too literal and at the same time creates curiosity. One good example is Plax mouth rinse—the name is pronounced plaque (the stuff the dentist cleans off your teeth), perfectly describing what it does, and because it is not a real word it is an allowable trademark.

Trademark Your Product

Once you have a brand name and distinguishing artwork ready, you should file a trademark application. Federal registration helps to legally protect your product image. The name of your trademarked product will be followed by ™ or ® to show that it is a registered name. Again, you may need a patent attorney to assure proper filing of your registration since a study in 2002 showed that of the 4,000 trademark applications filed in one week, 52 percent contained errors in the design codes assigned by USPTO.

Many people confuse patents, copyrights, trademarks, and trade secrets. While each one protects intellectual property, they are different and serve an exact purpose. A patent grants property rights to the inventor for a specific number of years. A copyright protects literary, dramatic, artistic, and musical intellectual works. A trademark is a name, logo, word, or symbol that is used on a product to distinguish it from others. A trade secret is the recipe or method that is used to make a product. The person who developed it chose to keep the formula a secret instead of patenting it. Coke has all four—a patent, a trademark, a copyright, and a trade secret.

Now is a good time to reserve a domain name on the Internet. Your domain name should match the name of your product or relate to your product. It should also be easy to remember. More and more marketers are using the Internet combined with streaming video to help market and sell their products.

If you plan to market and sell your product yourself, you will need to obtain a Universal Product Code (UPC) for your product. Nearly everything has a UPC printed on it somewhere. The 12-digit Universal Product Code is used to allow items to be scanned by an electronic "eye" and facilitate faster checkout at store registers. It also helps a retailer or distributor keep track of inventory. GS1, formerly known as the Uniform Code Council, is the U.S. numbering organization that manages and administers barcodes. You may find more information about how to get a UPC on GS1's Web site: www.gs1.org.

Chapter Three

Marketing Your Product

Thomas Edison clearly understood that the business of selling a product is very different from the business of creating a product. The process of selling the product involves a lot of marketing savvy to help create a strong presence to attract a consumer to buy it. Remember, your product is competing with thousands of other products for the attention of a consumer, so you need to make your product stand out by using clever advertising that distinguishes your product and provides marketing leverage.

Secret #15: Use the Right Sales and Marketing Channels

Timing is also important when launching a product. Knowing that it takes years for a patent to be issued, you may want to start marketing your invention as soon as you file your patent application. You will probably already have your business plan with a strong marketing strategy mapped out at this point. Once your product has been created and is ready for the market, you will need to explore the avenues for getting your product into the marketplace. Here is a list of various marketing channels you should consider:

TV home shopping such as QVC and HSN;

TV advertising (either a "short form" 2-minute commercial or a "long-form" 30-minute infomercial);

Mail-order catalogues (including SkyMall);

Direct selling organizations (such as Avon);

Individual direct selling (such as trade shows, community festivals, and mall kiosks);

Retail stores (large stores such as Wal-Mart or small stores such as local gift shops);

Printed ads;

Internet;

Press releases.

Each of these marketing channels presents advantages and disadvantages. You will need to determine the best way to gain access to these channels in order to achieve success. Some products work better in certain channels; for example, cooking items that need to be demonstrated sell very well on TV.

Your marketing method should fit the product's design. Not all products are meant for TV advertising. If your product is something that requires touch, you have to get the product into the person's hands where they can feel it. You can't feel a product on television. I didn't realize how important it would be for the product to be demonstrated hands-on when I created an infomercial for my Nukkles massager. The infomercial cost a lot of money and the product did not perform well on TV. I took Nukkles to a fitness industry tradeshow and sold $7,200 worth of product. Then, I got the idea of taking them to a local mall kiosk where a vendor could give a demonstration and allow the customer to try them out. It worked! We sold over 1 million units during the two-month holiday season.

Kiosks

Before spending a ton of money on printed material or an infomercial for direct response TV, you may want to test the marketability of your product at a local mall kiosk. These carts are not that expensive to set up and manage, and most do not require a long-term contract. With the exception of October, November, and December you can usually get a short-term lease for one month at a time, and a few leasing managers may even allow you to test your product for a weekend. The cost to rent a mall kiosk during non-holiday time ranges between $1,500 to $2,500 per month depending on the city and average income. During the holiday season the cost may increase to $15,000 to $25,000 per month, but the average cart operator can generate sales that range from $50,000 to $250,000 if they have the right product. Mall carts also work great for products that have to be demonstrated.

The corridors and common areas of the malls were empty until about twenty years ago when a Frank Blumer and Ron Yoder of American Home Products started selling flying toy airplanes that loop back to the sender. The only retailers in malls at that time were anchors (department stores typically) and in-lines (the stores between anchors). Blumer and Yoder saw the perfect opportunity to demonstrate their airplanes by roping off an area in the mall hallway and vending directly to the passers-by, but getting the leasing managers to agree was not easy. They thought it was strange that a vendor wanted to lease space in the hallway, but eventually a few malls saw the potential and allowed them to set up for a monthly fee. These two entrepreneurs sold millions of toy airplanes, in spite of the protest from in-line toy retailers.

When malls started leasing smaller spaces, one space was not adequate for properly demonstrating the flying toy airplanes. Blumer and Yoder were unable to continue to make a profit when they needed to rent three spaces, so they launched another toy called the bungee ball that required only one space or kiosk to demonstrate. The bungee ball caught on quickly and bounced back generous revenues for American Home Products. Today, mall hallways are overflowing with kiosks that bring increasing revenues to the mall.

Infomercials and TV Ads

Television advertising is a major way to reach a large sect of potential customers nationwide A TV ad lasting less than one minute may be produced for under $10,000, but you will still need to pay for airtime. A short commercial usually asks the viewer to go to the store to purchase the product. Direct response marketing always seeks a call-to-action such as "to order now call 1-800-PUR-CHASE" or call for more information. Direct response TV business generates billions of dollars in sales each year, but the cost of producing an infomercial is not cheap! The cost to produce a typical low-end infomercial (30-minute segment) is $200,000. The price could be as much as $1 million for a high-end production. There are two media categories of direct response TV: the test and the roll-out. Once you have the infomercial produced, you will test it on local broadcast and national cable. If the test is successful, the roll-out process begins.

The cost of producing a short-form commercial (less than 2 minutes) costs $20,000 to $100,000. The cost for producing an infomercial or short-form commercial depends upon location shoot, background music, scripting, studio time, special effects, re-shooting, editing, overtime, and the acting talent. You may want to hire a low-cost spokesperson or celebrity to endorse it on an infomercial. Even though it will cost more, it may be well worth it.

Secret #16: Celebrity Endorsements

If I told you to take your favorite grill and cut the front legs off so the grease would run onto the floor you would probably ask, "Why should I break my grill?" A man named Michael Boehm thought it would be a good idea. He thought that since people were concerned about reducing fat in their diet but really did not want to give up grilling burgers and chicken, he would invent a product that would allow people to have their grilled meat and eat it too. He angled the grill so the fat would run off the grill.

Boehm also thought the product would sell very well on TV if George Foreman was hired as the spokesperson. Use a retired boxer to sell a broken grill? Yes, he was 100% right; over 70 million George Foreman Lean, Mean Fat-Reducing Grilling Machines have been sold since 1995, and they keep selling. One thing Michael Boehm didn't consider was the amount of success his product would have. Initially he didn't manufacture enough products to meet consumer demand, and sold out before new ones could be made and put on the store shelves. Therefore, he did not make nearly what he could have for such a clever invention.

Three years ago Jack Lalanne, a 1950s TV fitness personality, celebrated his 90th birthday and the success of selling over 2 million Jack Lalanne Power Juicers. Jack claims his long healthy life is from his many years of drinking fresh squeezed juice. Jack Lalanne is the perfect celebrity for this product: his honest belief, wit, and humor combined with the fact that he speaks to an aging population that wants to life forever make for a powerful product pitch. I am very fortunate to have met Jack on several occasions and I have to say, he really walks his talk.

Studies show that only 25 percent of the American population ever buys a product from a television offer. Any product appearing on TV should have a very broad appeal since television has a massive and diverse audience. There is a lot of competition for the viewer's money. The biggest sellers in TV ads are those which have sex appeal, offer financial opportunities, or involve self/personal improvement. The top rated categories are fitness and exercise equipment, housewares and kitchen gadgets, self-improvement or weight-loss products, cosmetics, health and beauty aids, tools, car care, money-making business opportunities, golf and fishing equipment, and music and video products. If your product is one that is right for TV advertising, you will want to look into infomercials.

To determine if your product will do well as an infomercial, ask these questions:

Is my product unique? Does it elicit a response?

Can my product be presented in a visually compelling manner?

Does the product have mass appeal or does it have a niche market?

Does the product solve a problem? Fulfill a need? Make life easier?

Is the product offered at a price that helps the customer feel they are getting a value deal?

Does the product have multiple uses?

Does your product category have a proven success rate?

Is the product driven by a dynamic, super-energetic individual or popular personality?

Does the product meet the 5:1 ratio of cost of goods?

A product advertised in a short 1-2 minute spot should retail between $9.95 and $19.95, but a product offered on a half-hour infomercial should retail for at least $59.95 and up to $1,200, with a payment trial plan.

If you can answer yes to most of these questions, your product may be a great candidate for direct response TV.

One half-hour block of national, cable, or local broadcast time ranges between $800 and $30,000. Purchasing air time is best left to professionals who have long-standing relationships with key contacts and know where to buy and when. They will have access to data that contains TV product marketing and success histories, which are important when seeking the best rates and time slots. An agent that buys for many clients also has tremendous clout with station managers. The commission you pay to a media buyer could mean the difference in obtaining any old spot and obtaining a spot where your product will succeed.

You should allot 5-7 percent of your total campaign budget (not including postage) for fulfilling orders generated by the infomercial. If you are asking the customer to order by phone, you will need a call center and people to answer the phone calls. Depending upon how much data you want to capture from your customers, each phone deal can cost between $1.50 and $5.00 per call. Plus, you will need a way to process credit card payments, which will take approximately 3 to 4 percent of each sale.

It is impossible to predict how many people will call in response to your infomercial, but the industry average demonstrates $2.50 in sales for every dollar spent on the total project. It's not unusual for a successful marketer to go through several rounds of testing and tweaking before producing a successful direct response TV campaign. You may invest more than $1 million before you ever see one dollar of return.

Since studies show that only one of out twenty infomercials do well enough to generate a worthwhile profit, you should not plan to invest any money in direct response TV that you can't afford to lose. If you can't afford to lose that much money, there are other options. You may find a wealthy partner to go in with you, or you may license your product to a television marketing company. If you decide to license, I suggest you watch infomercials on television and make note of the ones that interest you or that seem to fit your product. Three-fourths of the infomercials you see on TV are produced by one of the five or six major infomercial companies. John or Clare Kogler at Jordan Whitney, Inc. in Tustin, California, can tell you who is behind the commercials. You may visit their Web site at www.jwgreensheet.com/pconsult.htm or phone them at (714) 832-2432. Get a recent copy of *DRTV Response Magazine* (www.responsemagazine.com/responsemag/). The members of the Direct Marketing Assocation DRTV Council www.directmag.com/news/dma-council-officers/index.html are able to give advice and help inventors with questions about direct market television.

Printed Ads

There is an industry term in advertising known as "cost per thousand" (CPM). One unit refers to a group of 1,000 people. For example, if your printed ad cost you $5,000 and reaches 100,000 readers, your CPM or cost to reach a thousand people is $50. This includes all media whether it is printed, verbal (radio), visual (TV), or by mail. Newspaper ads are the least expensive way to advertise and may have a CPM as low as two dollars, but it is hard to know how large a market actually sees them. However, magazines with a specialty such as *Tennis* might be a better place to promote the new tennis racket you've invented because the market is already targeted to those who are interested in tennis rackets.

The cost of a direct marketing campaign seems expensive, but if you calculate the number of households you reach with a TV ad, it is actually economical. Your CPM will be less than a dime per household. In that light, it could be considered an investment.

Mail-Order Catalogues

Mail-order catalogues are still popular today and have numerous categories that appeal to target markets such as apparel and accessories, computers, home and garden, arts and crafts, consumer electronics, gifts, arts and entertainment, food and gourmet, toys and hobbies, books, music and film, health and personal care, and business. And don't forget *SkyMall* for the frequent flyers.

Direct Selling Organizations

Door to door is not as popular as it once was because people are wary about opening their door to strangers. One way to reach this type of customer is by placing your product in a product catalogue that is distributed among friends who have parties. Avon, Pampered Chef, Tupperware, and other direct selling companies use representatives who have built a relationship with the client as they are introduced to friends of friends. Their clientele is both repeat and new customers.

Individual Direct Selling

This category includes tradeshows, community fairs and festivals, expos, mall carts conventions, and any other public gathering where vendors are allowed to market their wares. Impulse buying can be triggered when someone sees your new product being demonstrated.

Radio Ads

Radio advertising has a low CPM and producing a radio spot is much less expensive than a TV ad. The disadvantage of radio ads is that the customer who is interested in your product may not remember the phone number or Web site address if they can't write it down right away. They are not likely to pull off the road if they are driving to jot down the information; a newspaper or magazine ad can be clipped for future reference. Once the radio spot is heard, it may be forgotten.

Press Releases and Articles

A press release is an article about your product that is written as news, usually one page typed. Getting your press release published by a newspaper depends upon the quality of the writing and the newspaper's agenda. If you have trouble getting your press release or article published by the newspaper, you may want to seek publication in an industry trade magazine. The circulation may be less than 100,000 subscribers, but your news will be read by those in your target market.

Regardless of where you publish, you will want to use a standard format which makes it easier to read by editors. Your well-written piece will:

> Include your contact information at the top and bottom of the document;
>
> Have a title that catches the attention of the reader who would most benefit from your product;

Offer the product/service benefits in the title and first two sentences;

Be typed double-spaced;

Not break paragraphs if it goes to a second page (ideally a press release should be only one page);

Be free of typographical errors;

End with ### or ★★★ to indicate the end of the document.

Even though you are promoting your product or service and you are making a sales pitch, you must make it sound like news.

Direct Mail to the Home

Getting into the home of a potential customer is best done through direct mail marketing. Unlike telephone solicitation, which has been made more difficult by the "no call" legislation, printed mail is still an option for advertising your product. Try postcards as they work like little billboards and you can drive people to your Web site.

Retail Stores

This includes large department stores such as Wal-Mart and small stores such as local gift shops. When you have a finished, packaged product ready for retail you can contact the buyer or the retail store, or you can use a representative who advocates several different products for a retail store. I use reps for the major chains and go direct to the smaller chains.

Internet

If you do not have enough money for a large-budget marketing campaign and can't raise funds to get your product to the market, you may try Internet marketing. It is less costly than direct TV ads, and many products do well through the Internet.

The Internet is one of the greatest inventions of all times. Through Web sites and email marketing we can network, advertise, operate a business, and keep in touch with our customers. Internet marketing differs from traditional print advertising and person-to-person marketing, and there are quite a few strategies to reach customers and get them to buy online.

E-commerce newsletters allow us to develop relationships with our clients and let them know when we invent something they may be interested in. You may start with just a few names and email addresses asking those folks to forward your newsletter (it should have a link to a permission-based sign-up form) to their friends and urge them to sign up. Before long you have a new client base to market your product. You will not be spamming since everyone on your list has given you permission to email them.

The more traffic your Web site receives, the more sales you are likely to have. One way to increase traffic to your site is through search engine optimization (SEO). This means your Web site is clear of any obstacles that would deflect traffic and has plenty of keywords and metatags to attract search engines. I suggest you hire someone who has experience in this area to provide SEO for your site.

Another method of bringing more traffic to your site is by using pay per click programs where you can create your own ads and choose keywords that, when searched for, will bring up your ad. Both Yahoo and Google have these programs available. You pay only when someone clicks on your ad. The rate may be pennies per click or hundreds of dollars per click. The more you pay per click the higher your ad appears in the search engine results.

Google AdSense is a two-way program that allows you to generate revenue by placing other people's text or image ads on the pages on your Web site and in turn have your ad appear on the pages of sites related to your product or service. You can filter out your competitor's ads so they do not appear on your Web site. AdSense puts your ad all over the Web and creates links back to your site, which increases your ranking in the search engines. These are not pop-up ads that open a box or new screen while you are trying to read the text of a Web page, where the hyped up ad remains and may be hard to close. Pop-up ads are annoying and along with spamming, they are the worst marketing schemes available.

Banner ads are usually animated, colorful graphics that go across the top of a Web page. Skyscraper ads are similar but they run vertically on the left or right side of a page. Both ads link to another Web site (yours) and entice the visitor to leave the page they are on to see what the other site has to offer. These are losing popularity because people don't enjoy being distracted by the movement. Ads that don't move or talk are more effective.

Online magazines offer ad space for a fee (depending upon the magazine's popularity and traffic) and may appear for the entire time the current issue of the magazine is displayed. By selecting a magazine that relates to your product you are able to direct your ad to your target market rather than to a wide general audience.

Online city guides or industry directories such as Hellometro.com and CitySearch.com are Yellow Pages on the Internet instead of in print. They feature businesses in a particular area. For example, if someone in Atlanta is looking for a toy store, they may use the search engine to browse the online guide and find all the toy stores within a 10-mile radius of their zip code. In addition to the general listing, which is often free, you may purchase an ad for a related page.

Web site "lead generation" provides you with qualified leads at a set price per lead. These are collected by telemarketing companies who do cold calling to find potential clients for you. Lead generation services guarantee a particular level of return on investment and free your staff for other marketing efforts.

There are other types of Internet marketing, but one thing they all lack is personal interaction with the potential client. If you utilize some type of tracking software to record the email address of your Web site visitors, you may get a second chance to persuade a customer to buy. At least offer them the opportunity to become a subscriber to your permission-based newsletter.

You have to use words and pictures to sell your product when someone comes to your Web site. Your Web site also needs to have a call to action: ask the customer to buy something, or at least click on a link to get more information. Make sure you have an easy payment system in place to handle credit card purchases.

No matter what marketing effort you choose, make sure it works for your product and brings enough sales to cover the ad costs and make a profit. Otherwise, it doesn't warrant your time or money and you should find another way to market the product.

Chapter Four

Steering Clear of Invention Scams

When I came up with the idea for the Ab Roller, I decided to stay involved in the project every step of the way and take my invention all the way to the marketplace. I was involved in the entire process: designing, prototyping, branding, packaging, manufacturing, and marketing. I had an infomercial produced and aired the product on live television via the QVC Home Shopping cable channel. I became familiar with every aspect of the business of invention from the creation of an idea to the marketing of the product. I also learned a lot about protecting my intellectual property rights. Not only do you have to protect your product with patents and trademarks but you also have to protect your product from all the scam companies out there that want to make money off your idea before you ever make a penny.

How to Tell if an Invention Promotion Company Is a Scam

While it's tempting to start the process by looking for a company to promote or license your idea, scam companies are as common as inventions themselves and have robbed inventors of millions of dollars of their hard-earned money. There are several different types of scam companies out there. A few of them are actually legitimate marketing companies who are legally trying to sell you patent and trademark servic-es you do not need.

There are a few things about marketing and invention promotion companies that distinguish them from legitimate companies who truly want to work in your best interest. I will describe the frequently used tactics of these scam companies in nine scenarios. Not only will you be able to instantly recognize a scammer and save yourself a fortune, you may also help put those companies out of business for good.

Scam #1: We Do It All for You

Many scam companies offer to do everything for you from beginning to end. All you have to do is come up with an idea and present it to them. I wish I could offer a list of legitimate companies that will take an idea from conception to the marketplace, but unfor-tunately there is no one-stop shopping when it comes to inventions. Any company that offers to "do it all for you" and requires that you do nothing but pay them money is a scam. Invention is hard work and you are in the driver's seat when it comes to making decisions. However, there are some legitimate a la carte services that connect you with independent vendors who specialize in the different jobs you will need to perform along the way. For example, you may need to hire an expert to make a functioning prototype, or you may need to hire a qualified attorney to draft the patent application. You must shop around and compare prices for the services you need to complete the invention process.

Scam #2: Free Inventor's Kit

Some scam companies will place professional and appealing ads on radio, on TV, and in magazines to hook you. Some of them offer a FREE inventor's kit. First you call and receive in the mail a kit with a nondisclosure form and an offer to submit your invention for a free evaluation. Any company that asks you to submit your ideas without a signed contract or nondisclosure agreement should be avoided. Once you send in your idea, you will be contacted by a person from the scam company informing you that your idea is a good one and stating that you can probably get a patent. Not realizing that these companies tell every inventor that their idea is a good one that can be patented, you get all excited about the possibilities.

That is until they explain to you that in order to make sure you are not infringing upon someone else's patent they must first perform a patent search, which will cost you between $1,000 and $3,000. Don't be fooled: most legitimate patent attorneys charge $750 to $1,000 for a patent search.

Scam #3: Upfront Fees

Scam companies charge upfront fees that range from $5,000 to $40,000. If a salesperson or company wants money up front before they will do business with you, it's a sure warning sign that they are not on the up and up. If you can't pay the upfront fee, they may offer financing to help you get started. Once they get money from you, they will continue to ask for more, perhaps for an invention evaluation, research, marketing services, or a patent report. In contrast, rather than ask for fees in advance, reputable licensing agents rely on royalties from the successful licensing of their clients' inventions. Therefore, they are very discriminating about the ideas or inventions they pursue.

Regarding the research and evaluations offered by fraudulent firms, many times the so-called research report is nothing more than a mass-produced computer generated generic report that looks legitimate to the untrained eye. Their marketing evaluation may not be an honest appraisal of the feasibility or market potentiality of your idea.

Scam #4: Nothing in Writing

Scam companies make a lot of verbal promises they can't and don't intend to keep. They may promise to do a patent search, but if they don't offer you a contract with a patentability opinion signed by a patent attorney or agent, you have nothing to stand on if they don't follow through. If a company will not show you copies of any documentation regarding your agreement with them, this should be a red flag. If a company is above board, they will provide answers to your questions in writing. To avoid being a victim of scam, DO NOT accept verbal promises. Before you pay any money or sign any document with a marketing and invention promotion company, get answers to your questions in writing and make sure the document is signed by a company official. By the same token, be very careful about what you sign with an invention promotion firm. Some will have you sign a "non-disclosure agreement" that may actually keep you from disclosing your idea even in seeking patent for your own product.

Scam #5: It's in the Mail

There are several ways that scam companies operate using mail. They may try to convince you that your idea will be safeguarded if you put your idea in writing and mail a copy to yourself and don't open it.

They claim the postmark date will prove your date of invention, but this is not true.

Speaking of mail, for a fee of $4,000 to $9,000 these scam companies may offer to send a direct mail marketing piece to manufacturers. By the time you realize you are not getting a response from their efforts, you will know that they did not send the mailing and you have been scammed. By then, it may be too late to do anything about it.

Scam #6: No References

Fraudulent invention promotion companies may claim to know, represent, or have special access to manufacturers interested in licensing your invention. Try to get at least five references from past clients the company has serviced and contact them before signing any agreement. If a company refuses to provide references, check with the Better Business Bureau or your state's Attorney General's Office to see if the firm is registered. Also get a list of the manufacturers the firm works with and ask the manufacturers for references.

Scam #7: False Guarantees

Unscrupulous invention promotion companies will offer a "money-back guarantee" in the event they don't get a patent for you. There is no guarantee that a useful patent can be obtained until the research is completed. You can be sure you won't get your money back. These companies assure all inventors that their ideas have excellent market potential. Additionally, the scammer may tell you that you should apply for a design patent. A design patent only covers the appearance of a manufactured item, not the invention itself. If the deal sounds too good to be true, it probably is!

Scam #8: No One's Home in the Office

Another tell-tale clue that the company is a scam is the fact that you can't reach the company or its salespeople by phone, and when you leave a message your calls are not returned. Initially, when they are trying to lure you in, they may talk to you in person, but once they get your money, you are not able to speak with a representative when you have a question or need further assistance. The scam company may not have an office. It may be operating from someone's basement.

Scam #9: Patent First, Ask Questions Later

If an invention promotion firm asks you to apply for a patent and use their attorney to draft the application, you should beware. They really don't care whether your invention will sell; they just want your money as soon as they can get it. With all the wolves out there, it is easy to fall prey to scam companies and unscrupulous patent attorneys who are more interested in taking your money before knowing if your product will even sell. Just ask a patent attorney if he will waive his fees if you agree to pay him double once you start making money with your invention. He will look at you like you're crazy! It is important to focus on inventions that will SELL, not just inventions that can be patented.

To protect yourself against scammers, ask the following questions and get the marketing or licensing company's response in writing.

1. In the past five years, how many inventions has this firm evaluated for commercial potential? How many were accepted? How many rejected? The acceptance rate is usually less than 5 percent for legitimate firms.

2. How many clients have received license agreements for their inventions as a direct result of working with this firm?

3. As a direct result in working with this firm, how many clients have made more from their invention than they paid in fees to the firm?

4. Has the firm operated under names other than its present one?

5. Has the firm done business in other states?

6. Ask the firm to provide names and addresses of the manufacturers they have been affiliated with in the past ten years.

7. How many clients has this firm worked with in the past five years?

8. Has the Federal Trade Commission, Better Business Bureau, Attorney General's Office, or any consumer protection agency ever investigated this firm? If so, for what reason and what were the results?

9. Is there an upfront fee? If so, how much and what does the fee provide? What is the total fee involved in getting a patent and licensing agreement for your invention? Will royalties need to be paid to the firm? If the search agent charges a fee considerably less than the average agent, this may be a ploy to lure in novice inventors. By the same token, if large fees are required up front or if they ask for a percentage of royalties, head for the door!

10. Who will select and pay the patent attorney to do the patent search, give an opinion on patentability, and prepare the patent application?

11. Ask for the names, addresses, and phone numbers of five clients in your local area that the firm has worked with. Get copies of all contracts and forms used.

12. Ask whether the firm provides a written marketing evaluation report (opinion of the potential success of your invention). The marketing evaluation report some firms offer is something you can do for yourself for half the price.

Use common sense in evaluating the written answers from the promotion, marketing, or licensing company. Watch out for any company that refuses to offer proof of their marketing success, rejection rates, and relationships with manufacturers. Beware of those who will not give you details of its criteria for product evaluation or qualification.

If you are suspicious of a company, it's better not to deal with them. Be sure to report the company to the director at the Office of Independent Inventor Programs at the United States Patent and Trademark Office (write to Box 24, Washington, D.C. 20231; telephone 703-306-5568; or e-mail them at independentinventor@uspto.gov).

For specific companies you need to steer clear of, go to the Federal Trade Commission's Web site at http://search.ftc.gov/ and type the word "invention" in the search window. This will give you public information in the form of press releases about invention scam companies.

Suppose you are already involved with a scam company. If an invention promotion firm has caused you financial loss by making false or fraudulent statements, or by omitting any required disclosure, you have the right to sue the company and recover your loss plus attorneys' fees. You should also know about the Federal Trade Commission and Project Mousetrap—a joined effort of law enforcement and federal regulators to battle invention fraud (www.ftc.gov/opa/1997/07/mouse.htm). Contact the FTC and explain your situation. You may be able to get some of your money or rights back and keep other inventors from falling prey to the same scam company.

The US Patent and Trademark Office has no civil authority to bring case action against invention promotion firms, but it will accept and post your complaint online. You will need to complete the Complaint Regarding Invention Promoter Form, at www.uspto.gov/web/forms/2048.pdf. The Patent Office also has a public forum that lists complaints filed about invention promotion firms and the subsequent responses from the firms (www.uspto.gov/web/offices/com/iip/complaints.htm). Complaints about Advent Product Development; American Idea Management; American Inventors Corp.; Concept Network; Davison & Associates Inc.; International Product Design; International Product Design Inc; Invent-Tech; Invention Consultants, U.S.A; Invention Technology Co.; IP&R; National Invention Services, Inc.; New Product Advisory Group; New Product Consultants; New Product Consulting Corp.; New Products of America; and Patent Trademark Institute of America have been reported at the time of the writing of this book.

In October 2003, inventor Alan Scoggins found an ad on the Internet for Davison & Associates-an invention promotion company in Pittsburgh, Pennsylvania.

The ad promised that the company would perform a research design and patent search, create presentation materials, make a production sample, file disclosure documentation, and obtain a production quote and executive summary. Scoggins contracted to have the company "do it all" for him, but after paying over $15,000 to Davison and seeing no results, he got suspicious. When he reviewed the documents the company had sent him, he found that Davison's agreement requesting a product review with another company had been backdated two years prior to his contract with Davison. Knowing that the request for product review had never been made, Scoggins filed a complaint with the Patent Office stating that Davison had misrepresented their services in exchange for his money. Unfortunately, this kind of complaint is common on the Patent Office Web site.

The National Congress of Inventor Organizations (www.inventionconvention.com/ncio) and the Bureau of Consumer Protection (www.ftc.gov/ftc/consumer.htm) are other resources to help you prevent and recover loss. National Inventor Fraud Center (NIFC) (www.inventorfraud.com) was founded by Michael S. Neustel, a registered patent attorney. His advice to those who have been scammed is to go through the following steps:

Try to speak to a representative of the company and verbally request a refund.

If your verbal request fails, send a written request directly to an officer of the invention promotion firm via certified mail (e.g., President, CEO).

If you are unable to receive a refund yourself, you may want to get a consumer law attorney to help you. You may also want to file a complaint with your State Attorney General (see www.inventorfraud.com/attorneygenerals.htm for contact information for all 50 states). Also contact the federal Trade Commission and the USPTO.

To avoid being a victim of scam, *learn as much as you can about the invention process* and *watch for the scam warning signs*. Stay in control of the process through each step on the way to the market. *Hold off on getting a patent* until you are sure you have developed a product that can be *sold for a profit*.

Conclusion

Golden Rules for Riches

Perfect products make millions of dollars for everyone involved, but not all products will be perfect ones and many will never make it at all. Remember that failing is NOT a bad thing. Thomas Edison failed over 1,000 times while trying to invent the light bulb. He was not dismayed. He discovered 1,000 ways not to make a light bulb!

Only 20 percent of the products that hit the market each year will still be around twelve months later. Even the big companies make blunders every now and then. Do you remember Crystal Pepsi, Pepsi AM, or Pepsi Blue? None of these products are on the market today because they were not successful with consumers.

There are other reasons why inventions don't make it. Products can be ahead of their time, can't meet budget, or aren't marketed properly. Perhaps they fail because the market place is simply not ready for them. That's not to say that they will not be successful products later on or that those who invented them will never have a successful product.

I have made millions from my inventions, but I have also had many products that have failed. Those cost me hundreds of thousands of dollars but I gained experience and an education with each one. The business of invention is about learning from your failures.

When I first started teaching aerobic exercise I discovered that people are unable to count seconds and the beat of their pulse at the same time. I created Pulse Lite that worked like a traffic light. It was to be displayed on the wall at the gym where everyone could see it. I had an engineer create a circuit that continuously cycled in ten-second intervals. After locating your pulse on your neck or wrist, you would look at the Pulse Lite. The yellow light was a signal to get ready, the green light was to start counting, and when the light turned red you stopped counting. My mistake was that I did not understand cost-to-retail ratio on selling price. I overdesigned the product and it ended up costing me $250 to make each one. My customers were only willing to pay $300 for the product. That was not enough sales margin to make a profit. I lost $56,000 of my own money and $120,000 belonging to my investors. With present technology, I can make the product for about $20 and sell it all day long for $89.95. I may try it again someday.

Nukkles is a great massage product, but I made a big mistake in marketing it. I lost more than $200,000 trying to sell it on TV. However, I discovered the mall cart business in the process. These "show-and-tell" products do well in kiosks.

If major companies and successful inventors make mistakes, surely you are entitled to make some of your own. Just remember to use them as learning experiences and never to judge yourself harshly. It's best to laugh at your mistakes.

To recap what I've covered in this book, I will share with you some reasons why your invention should succeed.

1. You researched the industry, stores, and companies likely to want your invention.
2. You have a quality product that works and delivers more than what customers expect.
3. You did a really good job on your patent research.
4. You have a detailed plan about how you will either license your invention or manufacture it yourself.
5. Your invention can be produced at a price that has a 5:1 profit ratio while remaining competitive in the marketplace.
6. You are able to network with industry professionals.

7. You persevere when told you can't do something. There is always a way if the product is a good one.

8. You take responsibility and manage each step of the invention process.

In closing, here are my Golden Rules for Riches:

1. Companies don't care about ideas; they care about products, so invent a product that appeals to a large market.
2. Focus on selling, not suing.
3. Hold off on getting a patent until you are sure you have developed a product that can be sold for a profit.
4. The 5:1 ratio is an ironclad rule of retail. Make sure your product can be made at a cost that is 4-5 times less than your selling price.
5. Never fall in love with your idea. Try not to be too emotionally attached to your invention.
6. Never spend a penny on a patent until you are sure the product is going to sell. Remember that 97 percent of all patents never make a profit.
7. Only create products that deliver more than the customer expected.

8. Spend enough time researching your idea to make sure it is not already out there.

9. Never Give Up!

10. Did I say Never Give Up? NEVER GIVE UP!

If at first you don't succeed, keep inventing until you do. The secrets you have learned in this book will save you thousands of dollars and hours of time. Use them wisely and someday you too may INVENT YOURSELF RICH.

Part Two

Perfect Product Test & Inventor's Logbook

Perfect Product Test

You will have many different ideas on the way to inventing yourself rich, so I have included a place for you to get them out of your head and onto paper. Once you can see all your ideas in front of you, you can then start to evaluate each idea based on what you have learned from my book.

Use the following pages to list all of your invention ideas. Use one page for each idea. Write the idea at the top and date it. Next answer all the questions on the page with a yes, uncertain, or no. Be very honest with yourself. Total your score for the idea and write it at the bottom of the page. I suggest you have five friends fill out the questionnaire also. This will give you a better idea of how others would respond to the idea if it were a product. The higher your idea scores, the better chance of success it has in the marketplace.

Once you have completed this with all your ideas, it is time to take action and choose the one that is most likely to succeed and move it into the Inventor's Logbook. Your other ideas are now out of your mind but still on paper so you can always go back to them in the future. Many times good ideas are ahead of their time and may have a greater chance of success in the future.

For comparison, here are the scores of a few products mentioned in this book:

AB Roller: 28

Breathe Right strips: 27

Nukkles Massager: 26

Nukkles Soap: 16

Pulse Lite heartrate monitor: 15

Idea #1

Give your idea a quick name for remembering:

Describe your idea:

Date:

Perfect Product Test

1.Is your idea for a product a consumable product?

NO UNCERTAIN YES

2. Can you manufacture the product for 20% of the Retail Price?

NO UNCERTAIN YES

3. Is your idea so different that people will be convinced to change what they are currently using for your new product idea?

NO UNCERTAIN YES

4. Does your idea have a mass market appeal?

   NO UNCERTAIN YES

5. Can a customer instantly understand what your product does or is used for?

   NO UNCERTAIN YES

6. Does your idea fall into an established category?

   NO UNCERTAIN YES

7. Will your idea qualify for a utility patent?

   NO UNCERTAIN YES

8. Can your product idea be demonstrated on TV?

   NO UNCERTAIN YES

9. Is your product more remarkable or different than anything out there?

   NO UNCERTAIN YES

10. Will your product deliver much more than the customer expected?

    NO UNCERTAIN YES

Score value: NO=1 UNCERTAIN=2 YES=3

If your score is below 15, kill the idea before it kills your wallet!

If your score is between 16 and 24 continue development for 30 days, then if there's no resolution, kill the project.

If your score is above 25, go for it.

Idea #2

Give your idea a quick name for remembering:

Describe your idea:

Date:

Perfect Product Test

1.Is your idea for a product a consumable product?

NO UNCERTAIN YES

2. Can you manufacture the product for 20% of the Retail Price?

NO UNCERTAIN YES

3. Is your idea so different that people will be convinced to change what they are currently using for your new product idea?

NO UNCERTAIN YES

4. Does your idea have a mass market appeal?

   NO UNCERTAIN YES

5. Can a customer instantly understand what your product does or is used for?

   NO UNCERTAIN YES

6. Does your idea fall into an established category?

   NO UNCERTAIN YES

7. Will your idea qualify for a utility patent?

   NO UNCERTAIN YES

8. Can your product idea be demonstrated on TV?

   NO UNCERTAIN YES

9. Is your product more remarkable or different than anything out there?

   NO UNCERTAIN YES

10. Will your product deliver much more than the customer expected?

    NO UNCERTAIN YES

Score value: NO=1 UNCERTAIN=2 YES=3

If your score is below 15, kill the idea before it kills your wallet!

If your score is between 16 and 24 continue development for 30 days, then if there's no resolution, kill the project.

If your score is above 25, go for it.

Idea #3

Give your idea a quick name for remembering:

Describe your idea:

Date:

Perfect Product Test

1.Is your idea for a product a consumable product?

NO UNCERTAIN YES

2. Can you manufacture the product for 20% of the Retail Price?

NO UNCERTAIN YES

3. Is your idea so different that people will be convinced to change what they are currently using for your new product idea?

NO UNCERTAIN YES

4. Does your idea have a mass market appeal?

   NO UNCERTAIN YES

5. Can a customer instantly understand what your product does or is used for?

   NO UNCERTAIN YES

6. Does your idea fall into an established category?

   NO UNCERTAIN YES

7. Will your idea qualify for a utility patent?

   NO UNCERTAIN YES

8. Can your product idea be demonstrated on TV?

   NO UNCERTAIN YES

9. Is your product more remarkable or different than anything out there?

   NO UNCERTAIN YES

10. Will your product deliver much more than the customer expected?

    NO UNCERTAIN YES

Score value: NO=1 UNCERTAIN=2 YES=3

If your score is below 15, kill the idea before it kills your wallet!

If your score is between 16 and 24 continue development for 30 days, then if there's no resolution, kill the project.

If your score is above 25, go for it.

Idea #4

Give your idea a quick name for remembering:

Describe your idea:

Date:

Perfect Product Test

1.Is your idea for a product a consumable product?

NO UNCERTAIN YES

2. Can you manufacture the product for 20% of the Retail Price?

NO UNCERTAIN YES

3. Is your idea so different that people will be convinced to change what they are currently using for your new product idea?

NO UNCERTAIN YES

4. Does your idea have a mass market appeal?

   NO UNCERTAIN YES

5. Can a customer instantly understand what your product does or is used for?

   NO UNCERTAIN YES

6. Does your idea fall into an established category?

   NO UNCERTAIN YES

7. Will your idea qualify for a utility patent?

   NO UNCERTAIN YES

8. Can your product idea be demonstrated on TV?

   NO UNCERTAIN YES

9. Is your product more remarkable or different than anything out there?

   NO UNCERTAIN YES

10. Will your product deliver much more than the customer expected?

    NO UNCERTAIN YES

Score value: NO=1 UNCERTAIN=2 YES=3

If your score is below 15, kill the idea before it kills your wallet!

If your score is between 16 and 24 continue development for 30 days, then if there's no resolution, kill the project.

If your score is above 25, go for it.

Idea #5

Give your idea a quick name for remembering:

Describe your idea:

Date:

Perfect Product Test

1.Is your idea for a product a consumable product?

NO UNCERTAIN YES

2. Can you manufacture the product for 20% of the Retail Price?

NO UNCERTAIN YES

3. Is your idea so different that people will be convinced to change what they are currently using for your new product idea?

NO UNCERTAIN YES

4. Does your idea have a mass market appeal?

   NO UNCERTAIN YES

5. Can a customer instantly understand what your product does or is used for?

   NO UNCERTAIN YES

6. Does your idea fall into an established category?

   NO UNCERTAIN YES

7. Will your idea qualify for a utility patent?

   NO UNCERTAIN YES

8. Can your product idea be demonstrated on TV?

   NO UNCERTAIN YES

9. Is your product more remarkable or different than anything out there?

   NO UNCERTAIN YES

10. Will your product deliver much more than the customer expected?

    NO UNCERTAIN YES

Score value: NO=1 UNCERTAIN=2 YES=3

If your score is below 15, kill the idea before it kills your wallet!

If your score is between 16 and 24 continue development for 30 days, then if there's no resolution, kill the project.

If your score is above 25, go for it.

Idea #6

Give your idea a quick name for remembering:

Describe your idea:

Date:

Perfect Product Test

1.Is your idea for a product a consumable product?

NO UNCERTAIN YES

2. Can you manufacture the product for 20% of the Retail Price?

NO UNCERTAIN YES

3. Is your idea so different that people will be convinced to change what they are currently using for your new product idea?

NO UNCERTAIN YES

4. Does your idea have a mass market appeal?

NO UNCERTAIN YES

5. Can a customer instantly understand what your product does or is used for?

NO UNCERTAIN YES

6. Does your idea fall into an established category?

NO UNCERTAIN YES

7. Will your idea qualify for a utility patent?

NO UNCERTAIN YES

8. Can your product idea be demonstrated on TV?

NO UNCERTAIN YES

9. Is your product more remarkable or different than anything out there?

NO UNCERTAIN YES

10. Will your product deliver much more than the customer expected?

NO UNCERTAIN YES

Score value: NO=1 UNCERTAIN=2 YES=3

If your score is below 15, kill the idea before it kills your wallet!

If your score is between 16 and 24 continue development for 30 days, then if there's no resolution, kill the project.

If your score is above 25, go for it.

Idea #7

Give your idea a quick name for remembering:

Describe your idea:

Date:

Perfect Product Test

1.Is your idea for a product a consumable product?

NO UNCERTAIN YES

2. Can you manufacture the product for 20% of the Retail Price?

NO UNCERTAIN YES

3. Is your idea so different that people will be convinced to change what they are currently using for your new product idea?

NO UNCERTAIN YES

4. Does your idea have a mass market appeal?

NO UNCERTAIN YES

5. Can a customer instantly understand what your product does or is used for?

NO UNCERTAIN YES

6. Does your idea fall into an established category?

NO UNCERTAIN YES

7. Will your idea qualify for a utility patent?

NO UNCERTAIN YES

8. Can your product idea be demonstrated on TV?

NO UNCERTAIN YES

9. Is your product more remarkable or different than anything out there?

NO UNCERTAIN YES

10. Will your product deliver much more than the customer expected?

NO UNCERTAIN YES

Score value: NO=1 UNCERTAIN=2 YES=3

If your score is below 15, kill the idea before it kills your wallet!

If your score is between 16 and 24 continue development for 30 days, then if there's no resolution, kill the project.

If your score is above 25, go for it.

Idea #8

Give your idea a quick name for remembering:

Describe your idea:

Date:

Perfect Product Test

1.Is your idea for a product a consumable product?

NO UNCERTAIN YES

2. Can you manufacture the product for 20% of the Retail Price?

NO UNCERTAIN YES

3. Is your idea so different that people will be convinced to change what they are currently using for your new product idea?

NO UNCERTAIN YES

4. Does your idea have a mass market appeal?

   NO UNCERTAIN YES

5. Can a customer instantly understand what your product does or is used for?

   NO UNCERTAIN YES

6. Does your idea fall into an established category?

   NO UNCERTAIN YES

7. Will your idea qualify for a utility patent?

   NO UNCERTAIN YES

8. Can your product idea be demonstrated on TV?

   NO UNCERTAIN YES

9. Is your product more remarkable or different than anything out there?

   NO UNCERTAIN YES

10. Will your product deliver much more than the customer expected?

    NO UNCERTAIN YES

Score value: NO=1 UNCERTAIN=2 YES=3

If your score is below 15, kill the idea before it kills your wallet!

If your score is between 16 and 24 continue development for 30 days, then if there's no resolution, kill the project.

If your score is above 25, go for it.

Idea #9

Give your idea a quick name for remembering:

Describe your idea:

Date:

Perfect Product Test

1.Is your idea for a product a consumable product?

NO UNCERTAIN YES

2. Can you manufacture the product for 20% of the Retail Price?

NO UNCERTAIN YES

3. Is your idea so different that people will be convinced to change what they are currently using for your new product idea?

NO UNCERTAIN YES

4. Does your idea have a mass market appeal?

   NO UNCERTAIN YES

5. Can a customer instantly understand what your product does or is used for?

   NO UNCERTAIN YES

6. Does your idea fall into an established category?

   NO UNCERTAIN YES

7. Will your idea qualify for a utility patent?

   NO UNCERTAIN YES

8. Can your product idea be demonstrated on TV?

   NO UNCERTAIN YES

9. Is your product more remarkable or different than anything out there?

   NO UNCERTAIN YES

10. Will your product deliver much more than the customer expected?

    NO UNCERTAIN YES

Score value: NO=1 UNCERTAIN=2 YES=3

If your score is below 15, kill the idea before it kills your wallet!

If your score is between 16 and 24 continue development for 30 days, then if there's no resolution, kill the project.

If your score is above 25, go for it.

Idea #10

Give your idea a quick name for remembering:

Describe your idea:

Date:

Perfect Product Test

1.Is your idea for a product a consumable product?

NO UNCERTAIN YES

2. Can you manufacture the product for 20% of the Retail Price?

NO UNCERTAIN YES

3. Is your idea so different that people will be convinced to change what they are currently using for your new product idea?

NO UNCERTAIN YES

4. Does your idea have a mass market appeal?

   NO UNCERTAIN YES

5. Can a customer instantly understand what your product does or is used for?

   NO UNCERTAIN YES

6. Does your idea fall into an established category?

   NO UNCERTAIN YES

7. Will your idea qualify for a utility patent?

   NO UNCERTAIN YES

8. Can your product idea be demonstrated on TV?

   NO UNCERTAIN YES

9. Is your product more remarkable or different than anything out there?

   NO UNCERTAIN YES

10. Will your product deliver much more than the customer expected?

    NO UNCERTAIN YES

Score value: NO=1 UNCERTAIN=2 YES=3

If your score is below 15, kill the idea before it kills your wallet!

If your score is between 16 and 24 continue development for 30 days, then if there's no resolution, kill the project.

If your score is above 25, go for it.

Inventor's Logbook

The Inventor's Logbook, similar to a diary, is a record of your ideas, inventions, patent process, experimentation, drawings, mileage log, notes, signatures, and other details. You will use it to document a chronological history throughout the patent process and show each step of the invention's development in a way that will provide proof of ownership in a court of law. Be diligent about keeping your Inventor's Logbook. Use it every time you work on your project.

Here are some Quick Tips for using the Inventor's Logbook:

1. Use permanent ink, not pencil.
2. If you make a mistake, mark through it with a single line, but do not erase data or tear pages out of the Logbook. Add your initials to anything you mark through, then enter the correct data nearby.
3. Write legibly.
4. Use a separate page for each day or event.
5. Start at the top of the first page and move down that page, then to the next page in forward succession. Do not skip around or leave blank pages. Instead, draw lines through any blank spaces or unused pages.

6. Make sure your Logbook includes the conception date of your invention, what problem it is supposed to solve, a description of how the product will look or work, how to manufacture it, and any other details about the invention.

7. Enter data in the logbook each time you work on your invention or talk to someone about it.

8. Sign and date each page.

9. Get signatures from each person you talk with about your invention and put a date on the page with the signature. Do not let anyone else write in your logbook except for their signature!

10. Make the entry brief, clear, and concise, but make sure you give enough details that anyone could understand the concept you are recording.

11. Consistently use headings such as: Observations, Concepts, Drawings, Results, Figures, Calculations, and References throughout the book.

12. If you have material such as printouts, photos, receipts, and quotes, make sure you permanently affix (tape, glue, or staple) them to a page in the proper location and chronological

sequence. Make sure you describe the attachment, and sign and date the page. If the material is too large to fit in the logbook (blueprints, engineered drawings), you may keep it in an additional record book, but make sure you mention the book and the attachment in the Inventor's Logbook.

13. When documenting patent activities, make sure you have a witness sign the entry stating that he or she saw or understood that the event took place.

14. Maintain consistent progress to show due diligence and that you are making a conscious effort to move forward with the invention. If you are inactive on the project for a while, note the reason why you were inactive (i.e., waiting on patent attorney to write the application).

If you are still unsure about what to write, see Chapter 2 of this book, *Secret #9: Protecting Yourself Without a Patent.*

In summary, a page in a logbook should contain the date, name of invention, subject of the entry, participants, signatures, and witnesses.

The following pages are an official Inventor's Logbook and may be used to record your ideas.

Subject of Entry ______________________________

Entered by ______________________ Date __________

Participants ______________________________

Witness Signature ______________________ Date __________

Witness Signature ______________________ Date __________

Subject of Entry ______________________________

Entered by ______________________ Date __________

Participants ______________________________

Witness Signature __________________ Date __________

Witness Signature __________________ Date __________

Subject of Entry ____________________

Entered by ____________________ Date __________

Participants ____________________

Witness Signature ____________________ Date __________

Witness Signature ____________________ Date __________

Subject of Entry ____________________

Entered by ____________________ Date ____________

Participants ____________________

Witness Signature ____________________ Date ____________

Witness Signature ____________________ Date ____________

Subject of Entry ______________________________

Entered by ______________________ Date __________

Participants ______________________________

Witness Signature ______________________ Date __________

Witness Signature ______________________ Date __________

Subject of Entry ______________________________

Entered by ______________________ Date ____________

Participants ______________________________

Witness Signature ______________________ Date ____________

Witness Signature ______________________ Date ____________

Subject of Entry ______________________________

Entered by ______________________ Date __________

Participants ______________________________

Witness Signature ______________________ Date __________

Witness Signature ______________________ Date __________

Subject of Entry ______________________________

Entered by ______________________ Date ____________

Participants ______________________________

Witness Signature __________________ Date ____________

Witness Signature __________________ Date ____________

Subject of Entry ______________________________

Entered by ______________________ Date ____________

Participants ______________________________

Witness Signature ______________________ Date ____________

Witness Signature ______________________ Date ____________

Subject of Entry ______

Entered by ______ Date ______

Participants ______

Witness Signature ______ Date ______

Witness Signature ______ Date ______

Subject of Entry ______________________________

Entered by ______________________ Date ____________

Participants ______________________________

Witness Signature ______________________ Date ____________

Witness Signature ______________________ Date ____________

Subject of Entry ______________________________

Entered by ______________________ Date __________

Participants ______________________________

Witness Signature ______________________ Date __________

Witness Signature ______________________ Date __________

Subject of Entry ______________________________

Entered by ______________________ Date __________

Participants ______________________________

Witness Signature __________________ Date __________

Witness Signature __________________ Date __________

Subject of Entry ___________________________

Entered by ___________________ Date ___________

Participants ___________________________

Witness Signature ___________________ Date ___________

Witness Signature ___________________ Date ___________

Subject of Entry ____________________

Entered by ____________________ Date ____________

Participants ____________________

Witness Signature ____________________ Date ____________

Witness Signature ____________________ Date ____________

Subject of Entry ____________________

Entered by ____________________ Date ____________

Participants ____________________

Witness Signature ____________________ Date ____________

Witness Signature ____________________ Date ____________

Subject of Entry ______________________________

Entered by ______________________ Date __________

Participants ______________________________

Witness Signature ______________________ Date __________

Witness Signature ______________________ Date __________

Subject of Entry ______________________________

Entered by ______________________ Date ____________

Participants ______________________________

Witness Signature ____________________ Date ____________

Witness Signature ____________________ Date ____________

Subject of Entry ______________________________

Entered by ______________________ Date ____________

Participants ______________________________

Witness Signature ______________________ Date ____________

Witness Signature ______________________ Date ____________

Subject of Entry ______________________________

Entered by ______________________ Date __________

Participants ______________________________

Witness Signature __________________ Date __________

Witness Signature __________________ Date __________

Subject of Entry ______________________________

Entered by ______________________ Date ____________

Participants ______________________________

Witness Signature ______________________ Date ____________

Witness Signature ______________________ Date ____________

Subject of Entry ______________________________

Entered by ______________________ Date ____________

Participants ______________________________

Witness Signature ______________________ Date ____________

Witness Signature ______________________ Date ____________

Subject of Entry ______________________________

Entered by ______________________ Date __________

Participants ______________________________

Witness Signature ______________________ Date __________

Witness Signature ______________________ Date __________

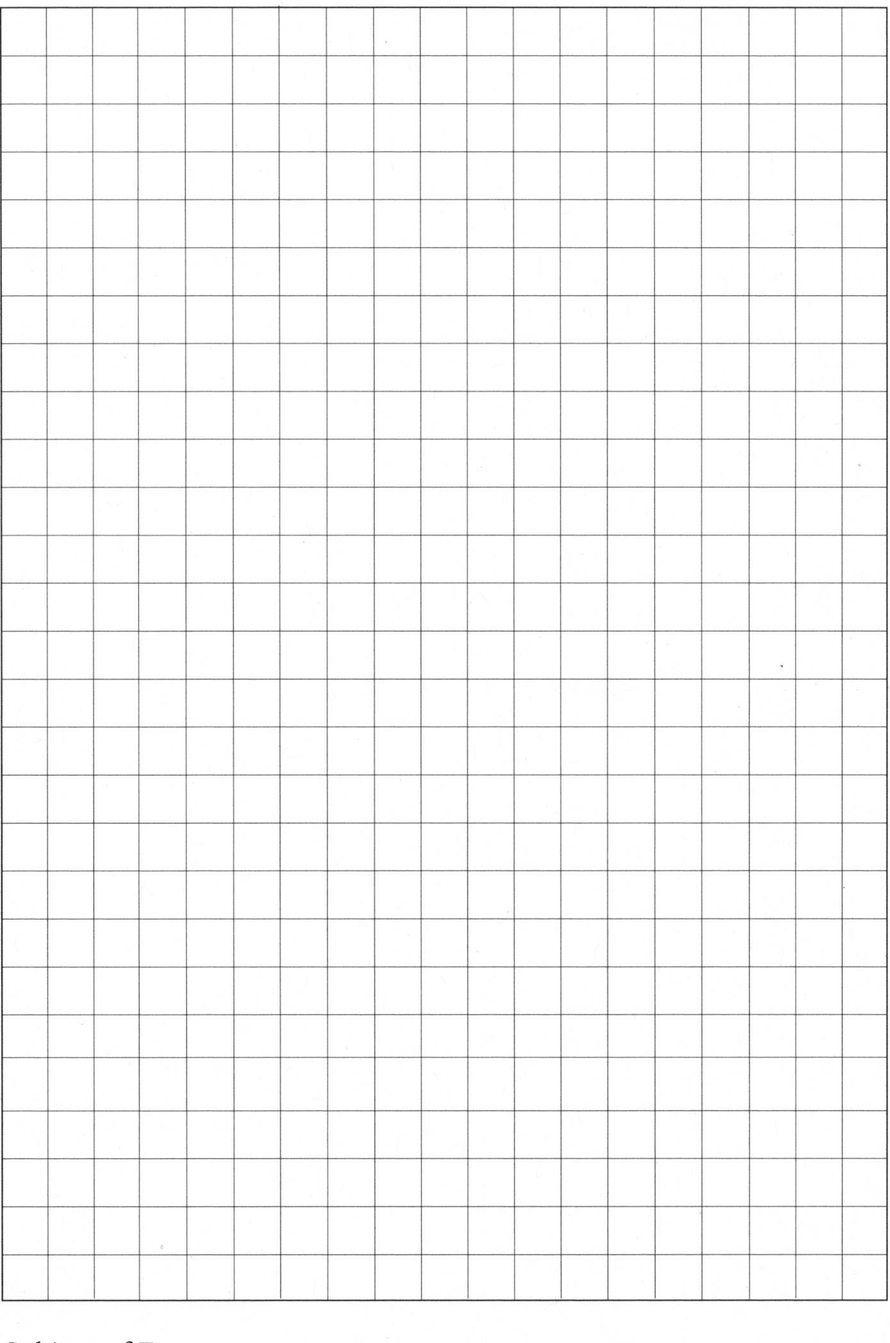

Subject of Entry ______________________________

Entered by ______________________ Date ____________

Participants ______________________________

Witness Signature ______________________ Date ____________

Witness Signature ______________________ Date ____________

Subject of Entry ______________________________

Entered by ______________________ Date ____________

Participants ______________________________

Witness Signature __________________ Date ____________

Witness Signature __________________ Date ____________

Subject of Entry ______________________________

Entered by ______________________ Date __________

Participants ______________________________

Witness Signature ______________________ Date __________

Witness Signature ______________________ Date __________

Subject of Entry ____________________

Entered by ____________________ Date ____________

Participants ____________________

Witness Signature ____________________ Date ____________

Witness Signature ____________________ Date ____________

Subject of Entry ______________________________

Entered by ______________________ Date ____________

Participants ______________________________

Witness Signature ______________________ Date ____________

Witness Signature ______________________ Date ____________

Subject of Entry ______________________________

Entered by ______________________ Date __________

Participants ______________________________

Witness Signature ______________________ Date __________

Witness Signature ______________________ Date __________

Subject of Entry ______

Entered by ______ Date ______

Participants ______

Witness Signature ______ Date ______

Witness Signature ______ Date ______

Subject of Entry ____________________

Entered by ____________________ Date ____________

Participants ____________________

Witness Signature ____________________ Date ____________

Witness Signature ____________________ Date ____________

Subject of Entry ______________________________

Entered by ______________________ Date __________

Participants ______________________________

Witness Signature ______________________ Date __________

Witness Signature ______________________ Date __________

Subject of Entry ______

Entered by ______ Date ______

Participants ______

Witness Signature ______ Date ______

Witness Signature ______ Date ______

Subject of Entry ______________________________

Entered by ______________________ Date ____________

Participants ______________________________

Witness Signature ______________________ Date ____________

Witness Signature ______________________ Date ____________

Subject of Entry ______________________________

Entered by ______________________ Date ____________

Participants ______________________________

Witness Signature ______________________ Date ____________

Witness Signature ______________________ Date ____________

Subject of Entry ______________________________

Entered by ______________________ Date ____________

Participants ______________________________

Witness Signature ______________________ Date ____________

Witness Signature ______________________ Date ____________

Subject of Entry ____________________

Entered by ____________________ Date __________

Participants ____________________

Witness Signature ____________________ Date __________

Witness Signature ____________________ Date __________

Subject of Entry ____________________

Entered by ____________________ Date ____________________

Participants ____________________

Witness Signature ____________________ Date ____________________

Witness Signature ____________________ Date ____________________

Subject of Entry ____________________

Entered by ____________________ Date __________

Participants ____________________

Witness Signature ____________________ Date __________

Witness Signature ____________________ Date __________

Subject of Entry ______________________________

Entered by ______________________ Date ____________

Participants ______________________________

Witness Signature ______________________ Date ____________

Witness Signature ______________________ Date ____________

Subject of Entry ______________________________

Entered by ______________________ Date ____________

Participants ______________________________

Witness Signature ____________________ Date ____________

Witness Signature ____________________ Date ____________

Subject of Entry ______________________________

Entered by ______________________ Date ____________

Participants ______________________________

Witness Signature ____________________ Date ____________

Witness Signature ____________________ Date ____________

Subject of Entry ______________________________

Entered by ______________________ Date ____________

Participants ______________________________

Witness Signature ____________________ Date ____________

Witness Signature ____________________ Date ____________

Subject of Entry ____________________

Entered by ____________________ Date ____________

Participants ____________________

Witness Signature ____________________ Date ____________

Witness Signature ____________________ Date ____________

Subject of Entry ______________________________

Entered by ______________________ Date ______________

Participants ______________________________

Witness Signature ______________________ Date ______________

Witness Signature ______________________ Date ______________

Subject of Entry ________________________________

Entered by ____________________ Date ____________

Participants ________________________________

Witness Signature ____________________ Date ____________

Witness Signature ____________________ Date ____________

Subject of Entry ____________________

Entered by ____________________ Date ____________

Participants ____________________

Witness Signature ____________________ Date ____________

Witness Signature ____________________ Date ____________

Subject of Entry ______________________________

Entered by ______________________ Date ____________

Participants ______________________________

Witness Signature ______________________ Date ____________

Witness Signature ______________________ Date ____________

Sample Confidentiality and Non-Disclosure Agreement

THIS CONFIDENTIALITY AND NON-DISCLOSURE AGREEMENT (the "Agreement") is entered into this ___ day of ____________, ____ (Year) between ___________________________ [your name] ("Inventor"), an individual, and ___________________ [Recipient's name] ("Recipient"), relating to certain terms and conditions that will govern the relationship between the above parties regarding the subject matter contained herein.

WHEREAS, Inventor has developed a proprietary product (the "Product") relating to [describe product very generally] ______________________________;

WHEREAS, in order to pursue a possible working relationship with Inventor with respect to the Product, Recipient is interested in receiving certain proprietary information from Inventor relating to the Product, including know-how, designs, and trade secrets which relate to the Product, including any samples thereof, demonstrations, reports, test data, photographs, patent applications, and sales, marketing, distribution and manufacturing processes (collectively, the "Proprietary Information"); and

WHEREAS, as an inducement for Inventor to provide Recipient with the Proprietary Information, Recipient has agreed to enter into this Agreement and to perform each and every one of the covenants herein, and understands any disclosure of Proprietary Information to others without the same type of restrictions set forth herein could be detrimental to Inventor and his affiliates.

NOW THEREFORE, in order to induce Inventor to disclose and Recipient to accept the Proprietary Information, each party agrees as follows:

1. Recipient shall receive in confidence the Proprietary Information as being proprietary and confidential (whether or not any specific Proprietary Information is explicitly indicated as such at the time of disclosure, unless it is expressly designated as non-confidential). As a condition to Recipient's being furnished any Proprietary Information, Recipient hereby acknowledges and agrees that Recipient will not challenge the assertion that Inventor is the only entity authorized to manufacture, market and sell the Product. Recipient acknowledges that the Proprietary Information is essential to Inventor's ability to market and distribute the Product. Recipient further acknowledges and agrees for the benefit of Inventor to treat the Proprietary Information in accordance with the provisions of this Agreement.

2. Recipient hereby covenants to refrain from disclosing any Proprietary Information, to any third parties, or using the Proprietary Information for any purpose whatsoever other than in considering a possible working relationship with Inventor.

3. Recipient will return to Inventor within five (5) days of a written demand, and in any event, within five (5)

days after termination of discussions regarding Recipient's participation with Inventor, all Proprietary Information in his/her/its possession, custody or control, including all copies of Proprietary Information and all copies of documents and work created by Recipient or his/her/its affiliates, agents or third parties relating to the Proprietary Information, and to thereafter not use any knowledge derived from such Proprietary Information or to take any action with respect to the Product.

4. The undersigned parties agree that the above-stated obligations shall not apply to any Proprietary Information which:

a) is demonstrated with reasonable written proof by Recipient to have been in his/her/its possession prior to receipt thereof from Inventor; or

b) is demonstrated with reasonable written proof by Recipient to have been received by Recipient in good faith from a third party not subject to a confidential obligation to the other party; or

c) is demonstrated with reasonable written proof by Recipient to have been publicly known at the time of receipt or has since become publicly known other than by breach of this Agreement.

5. Notwithstanding the provisions of paragraph 2 above, Recipient may disclose Proprietary Information if Recipient is required in the opinion of counsel to make such disclosure as a result of a court order, subpoena or

law, and not for commercial gain. Upon any such event, Recipient will give Inventor prompt written notice on Recipient's receipt of any such order or subpoena, and a reasonable opportunity (in the circumstances) shall be given to seek a protective order.

6. This Agreement shall remain in force for a period of 36 months and be binding on the shareholders, officers, directors, employees, affiliates, agents, advisors, successors and assigns of each party. Recipient agrees that it will not disclose any Proprietary Information whatsoever to third parties with whom Recipient shares an economic relationship without Inventor's express prior written consent and the receipt of express written assurances of such third party that they will adhere to the provisions of this Agreement.

7. This Agreement, including its existence, validity, construction and operational effect, shall be governed by the laws of the State of ____[your State]_____, in the United States of America. The parties hereto hereby consent to the jurisdiction of the courts of the State of [your State]________________, in the United States of America. The parties hereto also hereby agree that the venue for any dispute or action arising from this Agreement shall be in the courts of the State of [your State]__________________, in the United States of America.

IN WITNESS WHEREOF, the undersigned parties have caused this Agreement to be entered into and made effective on the date first described above.

Dated:____________________

"RECIPIENT" (sign below)

____

[Print name:]

Dated:____________________

Inventor (sign below)

____

[Print name:]

Resources

U.S. Patent and Trademark Office, Commissioner for Patents, P.O. Box 1450, Alexandria, VA 22313-1450. Visit www.uspto.gov or call toll-free 1-800-PTO-9199

For more information about how to invent, see www.si.edu/harcourt/nmah/lemel/edison/html/how_to_invent.html.

For information about the Disclosure Document Program, Provisional Applications or Non-provisional Applications call toll-free 1-800-PTO-9199. Every state has a Patent and Trademark Depository Library that maintains collections of current and previously-issued patents and Patent and Trademark reference materials. To order a copy of the American Inventors Protection Act, call toll-free 1-800-PTO-9199, or visit www.uspto.gov/web/offices/com/speeches/s1948gb1.pdf.

For more information about partnership, please see www.ezinearticles.com/?Pitfalls-of-Having-a-Partner-in-Your-Online-Business&id=81792.

For a list of State Attorneys General see www.inventorfraud.com/attorneygenerals.htm.

National Congress of Inventor Organizations (NCIO) offers free articles, information, resources, and *America's Inventor Online* magazine. To contact NCIO, call toll-free 1-877-807-4074, or visit www.inventionconvention.com/ncio.

For UPC Barcode information, see www.gs1.org.

Products to help with your invention process: logbooks, resource books, and reference material: http://inventorhelp.fatcow.com/store/.

United Inventors Association (UIA) offers free articles, information, resources, referrals to local support groups for inventors, and online copies of its newsletter. To contact UIA, call 1-585-359-9310, or visit www.uiausa.com or www.uiausa.org.

Resources for grants to help fund your invention:

http://grants.nih.gov/grants/policy/policy.htm

www.grants.gov

http://patapsco.nist.gov/ts_sbir/

www.grantstrategies.com/business.html

http://grant-rating.com

www.product-reviews.org/grants.html

For more information and resources on taking your idea from your mind to the market place visit my website, www.inventyourselfrich.com

Bibliography

"About the NIFC." National Inventor Fraud Center. 25 May 2006. www.inventorfraud.com.

Adler, Carlye. "A Killer Idea." *Fortune Small Business.* February 2003: 56+.

Allen, Robbie. Robbie's MIT Musings. 21 June 2006. www.rallenhome.com/blog/mit-sdm/2005/03/john-osher-inventor-of-crest-spin-brush.html.

Allentuck, Andrew. "Failed Inventions." *Costco Connection.* July 1997: 22+.

American Inventor. ABC-TV. 1 May 2006. abc.go.com/primetime/americaninventor/about.html

Bellis, Mary. "The Invention of Velcro." 2 May 2006. inventors.about.com/library/weekly/aa091297.htm.

Brain, Marshall. "How UPC Bar Codes Work." 29 June 2006. www.howstuffworks.com/upc.htm.

"Bruce Johnson: Inventor of Breathe Right Nasal Strips." CNS, Inc. 12 May 2006 www.cns.com/inventors/inventorBJohnson.cfm

"Bungee Ball Ketch-it." 26 June 2006. www.ketch-it.com/History/history.html.

"Business Studies Marketing." Bbc.co.uk. 28 June 2006. www.bbc.co.uk/schools/gcsebitesize/business/marketing/brandingandpackagingrev2.shtml.

"CNS, Inc. Reports Strong Second Quarter Revenue and Earnings Growth." 13 November 2006. www.cns.com/news/releasedetail.cfm?ReleaseID=217052.

Deangelo, Linda. "Gizmos & Gadgets." *Inventors' Digest*, July-August 2002: 19.

"Direct Response Product Criteria." 19 May, 2006. www.morganjames.com/downloads/rateyourproduct.pdf.

Dobkin, Jeffrey. "Strategies to Help You Get in the Door." *Inventors' Digest*, September-October 2001:11+.

"Don't Be a Lone Wolf, Hunt with the Pack." Wilywalnut.com. 19 May, 2006. www.wilywalnut.com/Thomas-Edison-Creative-Invention-Secret-10-Hunt-Pack.html.

Eisenberg, Howard M. "Patent Law You Can Use." 13 November 2006. www.yale.edu/ocr/invent_guidelines/provisionals.html.

"Facts for Consumers." Federal Trade Commission for the Consumer. 25 May 2006. Facts for Consumers. www.ftc.gov/bcp/conline/pubs/services/invent.htm.

"Fascinating Facts About the Invention of Beanie Babies by H Ty Warner in 1993." The Great Idea Finder. 29 May 2006. www.ideafinder.com/history/inventions/beanies.htm

"Fascinating Facts About the Invention of Liquid Paper by Bette Nesmith Graham in 1951." The Great Idea Finder. 12 May 2006. www.ideafinder.com/history/inventions/story046.htm.

"FTC/State 'Project Mousetrap" Snares Invention Promotion Industry." 16 May 2006. www.ftc.gov/opa/1997/07/mouse.htm.

Gibbs, Andy. "Managing the Invention Process." *Inventors' Digest*, May-June 2001: 30.

Gnass, Stephen Paul. "Do You Have a List of Legitimate Companies?" National Congress of Inventor's Organizations. http://inventionconvention.com/b-12/

———. The "First-to-Invent" Patent System and the Inventors Log Book. Invention Convention. http://inventionconvention.com/ncio/inventing101/001.html

Godin, Seth. *Purple Cow*. New York: Penguin Group, 2003.

Hall, Doug. Judge for the American Inventor. 27 April 2006. http://www.eurekaranch.com/eureka/default.asp.

Hayes, Joanne. "One in the Hand." *Inventors' Digest*, May-June 2001: 10+.

Hayes-Rines, Joanne. "Cutting Outside the Box." *Inventors' Digest*, October-November-December 2005: 39.

———. "Re-Pillable." *Inventors' Digest*. October-November-December 2005: 43.

Hendrickson, John. "An Inventor's Guide to Infomercials and Other Forms of Direct Response Marketing." *Inventors' Digest*. May-June 2001: 22+.

"How Do I Get My Product on QVC?" 13 November 2006. www.vendor.studiopark.com/howto.asp.

"How It Works." Invention Perfection. 25 May 2006. inventionperfection.com/_wsn/page5.html.

Iknoian, Therese. "Bowflex Inventor on to Next Project." Dosho.com. 19 May 2006. www.dosho.com/h_press_mar1703.html.

"Internet Marketing Strategy: What Can it Do for You?" About.com. 15 November 2006. marketing.about.com/cs/marketingjobs/a/aanet-marktingb.htm.

Invention Notebook Guidelines. 26 July 2006. www.bookfactory.com/special_info/invent_note-book_guidelines.html.

"Invention Promotion Firms." Federal Trade Commission. 25 May 2006. search.ftc.gov/query.html?rq=0&col=full&col=hsr&col=news&qt=+invention&charset=iso-8859-1.

"Inventor of the Week." Massachusetts Institute of Technology. 19 May 2006. web.mit.edu/invent/iow/i-archive-cp.html.

"Inventor Resources." United States Patent and Trademark Office. 25 May 2006. www.uspto.gov/web/offices/com/iip/complaints.htm.

"John Osher's 17 Mistakes for Small Business Start-ups." Accessed 20 November 2006. www.ociinc.biz/articles.html#17_things.

Kirkeby, Cynthia, and U.S. Patent Office. "Top 10 Inventor Scams." Classbrain.com. 24 May 2006. www.classbrain.com/artteenah/publish/article_101.shtml.

Koeppel, Peter. "Testing, Tweaking, Retesting, Tweaking." 30 June 2006. www.koeppeldirect.com/media-buying-direct-response-Sept2003.htm.

Lander, Jack. "How to Benefit from the Professionally Designed Prototype." *Inventors' Digest*, July-August 2002: 31+.

———. "The Experts Say . . ." *Inventors' Digest*. October-December 2005: 36.

"Lava Ice." Bodytimewellness.com. 16 May 2006. www.bodytimewellness.com/LavaIce_Instructions.pdf#search='lava%20ice'.

Lim, Peter. Ezinearticles.com. 20 November 2006. www.ezinearticles.com/?Pitfalls-of-Having-a-Partner-in-Your-Online-Business&id=81792.

"Mark Henricks Interviews John Osher Who Discusses the 17 Most Common Mistakes Startups Make, Along with 5 Must Dos to Win." Accessed 20 November 2006. www.ociinc.biz/ articles.html#17_things.

McPherson, Robert. "Licensee=Partner." *Inventors' Digest*, July-August 2002: 40+.

Monosoff, Tamara. "Top Six Mistakes Inventors Make." Entrepreneur.com. 10 April 2006. www.entrepreneur.com/article/0,4621,327122,00.ht ml

Neustel, Michael S. "Inventor fraud information." www.inventorfraud.com. 25 May, 2006. www.entre-preneur.com/article/0,4621,327122,00.html

Niemann, Paul. "The 7 Major Types of Media." *Inventors' Digest*. May-June 2003: 28+.

———. "The Top 10 Reasons Inventions Succeed." *Inventors' Digest*. July-August 2002: 16+.

Nipper, Stephen M. "Another Invention Promotion Company Story." The Invent Blog. 29 May 2006. nip.blogs.com/patent/2006/01/another_inventi.html

Oppedahl & Larson LLP. "What Things Cost." 24 May 2006. www.patents.com/cost.htm.

Parmelee, Anna. "So You Want to Advertise Your Product on Television?" *Inventors' Digest*. March-April 1996: 20+.

"Patent Costs." Beem Patent Law Firm. 20

November 2006. www.beemlaw.com/patent_costs.htm.

Pemberton, John. "How to Select a Good Patent Attorney." *Inventors' Digest*, October/November/December 2005: 19+.

"Proceeds Benefit NY Fire & Police." Epinions.com. 21 June 2006. www.epinions.com/ content_72584564356.

"Roll Up Bag Irons Out Wrinkle Problem." News USA.com. 29 May 2006. about.newsusa.com/article-site.asp?ArticleId=3784.

Scoggins, Alan. "Complaint Regarding Invention Promoter." 1 June 2006. www.uspto.gov/web/offices/com/iip/complaints_508/davison-scoggins.htm.

"SkyRoll, the Unique Roll-Up Luggage That Lets You 'Travel Outside the Box,' Now Available Nationally at All Men's Wearhouse Stores." Travelwriters.com. 29 May 2006. main.travelwriters.com/pressreleases/about/details.asp?pressreleaseID=211.

"Spotting Sweet-Sounding Promises of Fraudulent Invention Promotion Firms." Federal Trade Commission. 25 May 2006. www.ftc.gov/bcp/conline/pubs/alerts/invnalrt.htm.

"State Attorneys General." National Inventor Fraud Center. 29 May 2006. www.inventorfraud.com/attorneygenerals.htm.

Stringer, Kortney. "If at First You Do Succeed." *Wall Street Journal Online*. 12 July 2004. online.wsj.com/ad/article/fedex/SB108929852671458525.html?mod=sponsored_by_fedex.

Szot, Colleen. "The Eight Most Frequently Asked Questions." 30 June 2006. http://www.wonderful-writer.com/about/faqs.

Tanker, Nancy. "Bungee Business." Entrepreneur Profile. http://www.bungee-ball.com/Specialty-Retail-Rpt-Article.html.

"Toys and Spinning Brushes: How John Osher Found His Way to Profits." Wharton School of the University of Pennsylvania. 21 June 2006. knowledge.wharton.upenn.edu/index.cfm?fa=viewfeature&id=870.

"Travel Outside the Box." SkyRoll.com. 29 May 2006.

Tripp, Alan R. *Millions from the mind: How to turn inventions (yours or someone else's) into fortunes.* New York: Amacom, 1992.

Types of Internet Marketing. Podium Solutions. 15 November, 2006. www.podiumsolutions.co.uk/eBusiness-Strategy-Guide/types-of-internet-marketing.html.

"United States Census 2000." U.S. Census Bureau Web site. 26 June 2006. www.census.gov/main/www/cen2000.html.

United States Patent and Trademark Office. Scam Prevention brochure. Washington, D.C.: GPO 2006. www.inventorfraud.com/pto.pdf.

"USPTO to Give Patent Filers Accelerated Review Option." U.S. Patent and Trademark Office. 19 May 2006. www.uspto.gov.

Warren Dobkin, Jeffrey. "Getting Your Press Release into Print." *Inventors' Digest*. March-April 1996: 42+.

"What Are Patents, Trademarks, Copyrights and Trade Secrets?" *Inventors' Digest*, July-August 2002: 9+.

"What Can Be Patented?" *Inventor's Handbook* 19 May 2006. web.mit.edu/invent/h-main.html

Wilhite, Ann. "Adopting Love." *Topic Magazine*. 3 July 2006. www.webdelsol.com/Topic/articles/04/wilhite.html.

"You Can't Live Without What?" *Inventors' Digest*, May-June 2003: 7.

Yubas, Matt. "An easy to follow roadmap to get your invention into the market." 26 April 2006. www.mattyubas.com.

Acknowledgements

I want to thank the following people for making this book possible: Doug Hall, Michael Gerber, Seth Godin, Dr. Wayne Dyer, Zig Ziglar, Denis Waitley, Brian Tracy, Tony Robbins, Christopher Reeves, Jay Abraham, Dan Kennedy and many others whose work has given me the college education I never received.

Thanks to Yvonne Perry for her research and writing skills: without your help this book would not have happened.

Thanks to all my business partners for questioning my decisions and forcing me to look outside the box.

Thanks to all my family and friends who have given me advice and support during my crazy life.

Thanks to my wife Tanga for putting up with my many business ventures.

Thanks to my daughter Courtney for showing me what creativity really means.

Thanks to my Grandmother for giving me countless things to fix and repair.

Thanks to my Mother for giving me the gift of life and creativity.

Thanks to everyone involved in the making of this book.

Thanks in advance to all the idea people who have purchased this book.